CHARLES H. M. BECK
LOUIS NEAL IRWIN

THE
EVOLUTIONARY
IMPERATIVE

First edition copyright ©2016 by Charles H. M. Beck and Louis Neal Irwin by CCB
Publishing, British Columbia, Canada.

To order additional copies of this book, contact:
Xlibris
844-714-8691
www.Xlibris.com
Orders@Xlibris.com

ISBN: Softcover   978-1-6641-8675-0
         Hardcover   978-1-6641-8676-7
         EBook   978-1-6641-8674-3

Library of Congress Control Number: 2021915048

Print information available on the last page

Rev. date: 09/02/2021

CONTENTS

Preface to the Second Edition

Charles Beck, the first author of this book, lost a valiant battle to cancer just as the first edition was about to be published in electronic form five years ago. It thus fell to me as the second author to incorporate feedback and contemplate how the manuscript might be improved with the publication of a second edition and first hard copy version of our book.

The original goal of the work envisioned by Charles was to make a broad and well-documented argument for the pervasive influence of two physical principles—the Second Law of Thermodynamics and the Principle of Least Action—on the changes in nature that occur over time, from the atomic to the galactic level, and from the behavior of single cells to the attributes of entire societies. While originally advanced by the Finnish physicist, Arto Annila, and his colleagues in numerous technical papers, Charles wanted to transmit the essence of Annila's insights through a text more accessible to the general public.

I have retained that vision in this new edition, while streamlining the argument and reducing some superfluous detail in a quest for greater clarity. The resulting book, I am confident, is consistent with what Charles wanted to say, if perhaps less granular in detail than he would have provided on his own. Principle responsibility for authorship of the different chapters remains as stated in the preface to the first edition, but final responsibility for the entire text obviously now rests with me.

I am very grateful for the insights that Annila, through Charles, provoked in me. Charles honored me by inviting me to co-author what was originally intended as his book. In that spirit, this new edition is affectionately dedicated to his memory.

Louis Neal Irwin

Preface to the First Edition

As co-authors of this work, we share two convictions about the evolution of all things. The first is that change in the natural world—at every level, from the subatomic to the cosmic; and for all time, from the big bang to the end of the world—is driven by a set of fundamental and invariant principles. The second is that change occurs because it must; that it is propelled by the certainty of physical imperatives governing the nature of matter and energy.

One of us, Charles Beck, is a neuroscientist specializing in psychopharmacology who has focused on the parallels between biological and behavioral evolution. He was inspired early in his career by the excitement of realizing that the shaping effects of selective reinforcement on behavior in psychology have a compelling analogy if not homology with the operation of Darwinian natural selection. Just as biological traits are selected by differential survival of organisms, behaviors are selected for repetition by the reinforcing effects of rewarded behavior. For social organisms like ourselves, groups of individuals behave in an organized fashion that

itself evolves over time, in apparent response to underlying principles governing the group above and beyond the individual. Thus societies, cultures, and institutions have evolutionary trajectories impelled by underlying principles, as much as the individuals of whom they are composed.

The other, Louis Irwin, is an evolutionary neurobiologist and astrobiologist who long has sought a way to integrate the principles that apply to evolution across all dimensions of the physical and living world. In particular, he has been searching for an explanation of why evolution happens, absent any physical law that requires it to do so. While short-term changes are inevitable in any dynamic system, the fact that evolution has directionality over time—and more specifically, a trajectory that tends toward greater order and complexity overall—has lacked a convincing explanation.

Since at least the middle of the twentieth century—with Harold Blum's seminal work, *Time's Arrow and Evolution*—the question of change in the natural world has been framed in terms of how the flow of energy through a system affects the organization of that system. The inverse relationship between entropy and organization, hence the Second Law of Thermodynamics (SLT), has been seen as integral to any comprehensive understanding of evolution. Building on this observation, and adding to it the less well-known Principle of Least Action (PLA), Arto Annila and his colleagues have offered a formal, comprehensive explanation of evolution which has sought to explain both its inevitability and its directionality.

Inspired by Annila's ambitious attempt to explain the salient features of evolution in terms of underlying physical principles, Beck wrote an essay some years ago that extended the reach of Annila's concepts to the nature of behavioral, social, and cultural changes, and suggested the implications of unfettered energy consumption for the fate of the biosphere. Over time, Beck wrote, evolutionary changes at all levels consume ever growing amounts of energy, as organization becomes more complex and the role of information becomes increasingly integral to the system. In reviewing Beck's essay, Irwin saw the prospect of a unifying picture for why evolution happens that he had been seeking. Thus motivated by different but complimentary concerns, the two of us decided to undertake a collaborative effort to bring these insights to a general audience; and this work is the culmination of that effort. Beck provided the driving inspiration for the project, and was principal author of chapters 6, 8, and 9. Irwin was primarily responsible for chapters 1-5. Both of us contributed to chapters 7 and 10.

The many authors from whose work our thoughts derive are cited at relevant points throughout the text. We have further benefitted from the valuable feedback and commentaries of Caitlin Beck, Christine Beck, Kath Beck, Brian Irwin, Rowan Stewart, and Sean Stewart.

Charles Beck
Louis Irwin

1

Statement of the Model

Change happens. We live in a dynamic world, where nothing stays the same forever. Oceans come and go, mountains rise and fall, new species emerge, and existing ones die out. Living cells are changing constantly. The entire organisms of which they are a part survive for minutes to centuries, depending on their size and metabolic rates. On at least one planet, but surely on millions of others, life has arisen to accelerate energy consumption. A subset of living organisms has evolved beyond the level of single cells to become much larger in size and even more effective transducers of energy. Among them, the human species has evolved to the point of creating technology and cultural behaviors that transcend the immediate requirements for survival, greatly accelerating the consumption of energy and degrading the environment in the process. Why is this happening? Where is it leading? Is this an irreversible trend? These are pressing questions for all of us. They are inextricably linked to the answers to the questions framing the book title—Who are we? Where did we come from? Where are we going?

Ever since that point in space and time when our universe came into existence, entropy (or disorder) has been increasing, and energy has been dissipating through the increased dispersal of matter and the release of heat and electromagnetic radiation. Even though the universe as a whole has been unraveling, its granularity has increased. Clumps of matter and energy have appeared as local pockets of complexity, forming stars, planetary systems, and nonrandom geophysical features like glaciers and mountain ranges.

Thus, the world has evolved simultaneously in opposite directions at different levels of resolution. It has become more complicated locally, while the local pockets of complexity with their surroundings taken together, have become more disordered as a whole. Thus, while the Sun and the planets and moons that orbit around it constitute a pocket of "local" complexity, the energy given off by the Sun, the transformation of that energy on planetary bodies, or its radiation into space, and the gradual changes in angular velocity of the planets and their moons, increase the entropy of the Solar System as a whole. Both the increase in local complexity and the increase in entropy of the total system are manifestations of different but interacting physical principles. The objective of this book is to seek a unified set of principles that accounts for the paradox of a world in which local pockets of complexity are generated in the process of diminishing order in the universe as a whole.

Contemporary science has good models for *how* change occurs. In the physical world, evolutionary changes, like the formation of spiral galaxies or the drainage systems of rivers, follow passively from well-established physical laws. In the living world, natural selection along with several other concepts provide a satisfactory model for how biological evolution occurs. In the social sciences, overarching principles have been more elusive, but are gaining in explanatory power in some disciplines. For example, certain biases in human thinking have been found to be universal.

On the other hand, *why* change occurs is not immediately obvious, nor is the direction of evolution inherently apparent. And in particular, that change should most often lead to higher levels of organization and increasing complexity does not follow from any basic law of nature. The ability of some species of ants to engage in agricultural farming practices, like horticulture, husbandry and crop storage, for example, must have provided the selective advantage that explains their natural selection, but why natural selection should have led to such complicated behavior, as opposed to other, simpler adaptations, is not apparent. The purpose of this book is to summarize the argument—currently available only in highly technical forms—of the scientific principles that can explain why change occurs and what general direction it is most likely to take. We refer to our version of the explanation as the **energy dissipation model of the evolutionary imperative**. Beyond our explanation of the model, we explore where evolution is leading on its current trajectory for our planet. Finally, we will address the question of how we might meet the challenges posed by the inevitable consequences of this evolution.

1.1 Why Evolution Happens

The starting point for our model is the observation that inhomogeneities in the fabric of space-time, presumably emanating from quantum fluctuations at the onset of the big bang, combined with the fundamental forces of nature to create local concentrations of matter and potential energy. These pockets of concentrated matter and greater potential energy constitute gradients with their surroundings, exerting a pressure for the outward dispersal of matter and the flow of energy from a state of higher to lower potential. The resolution of this tension accounts for all the dynamic properties of the universe. The world changes, or evolves, as matter is rearranged and potential energy is converted into motion, heat, light, chemical bonds, biomass, electricity, or other forms of useful (work producing) or non-useful energy. The nature of those changes, however, is not random. Two laws of nature are particularly important in constraining how they occur. The **Second Law of Thermodynamics (SLT)** determines the direction those changes must take—mandating that energy be released when work is done spontaneously and disorder is increased, and that energy be consumed when order is increased and potential energy is raised. The **Principle of Least Action (PLA)** dictates the course those changes must follow, requiring the shortest path to be taken that consumes the least amount of energy for a given amount of work. Our premise is that all the evolutionary changes that occur, both in the abiotic and living world, are resolutions of potential energy channeled by the PLA in the net direction required by the SLT. While information has a place in the Second Law, bearing an inverse quantitative relationship to entropy, its expression in the abiotic world is essentially passive, manifested in the ordered arrangement of matter like the structure of crystals and the non-random aggregation of land masses (Fig. 1.1a).

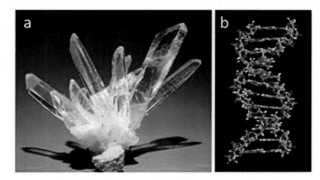

Fig. 1.1. Information in the non-living and living world. (a) An inert mineral such as quartz forms into a precise but relatively simple structural configuration according to laws of physical chemistry. (b) A molecule of DNA consists of constituents arranged in a highly ordered way, such that the precise configuration of its components conveys hereditary information essential for the proper functioning of the organism. *(a) Smithsonian Museum of Natural History; (b) Wikimedia Commons*

With the evolution of life, however, information assumes a more critical role. First, the function of a living cell or organism becomes much more dependent on its precise structure (order) than do dynamic processes in the non-living world (Fig. 1.1b). Second, the preservation of structure and function over time requires a coding mechanism that conveys information to subsequent generations of the replicated cell or organism. Since variation creeps into the informational code, with consequent changes in the form and function of the organism, a new process—natural selection—emerges to act as a filter for those variations that persist and those that do not. In the short run (proximately), the course of organic evolution is dictated primarily by natural selection. Over the long run (ultimately), evolution follows the path of Least Action (or Effort), in the direction required by the Second Law, because the evolutionary changes in accord with these laws are invariably more advantageous for survival. In other words, nature selects for survival the variant that consumes and dissipates energy most effectively and efficiently.

1.2 Where Evolution Leads

While the overall direction of change in the universe is toward increasing entropy, local reversals in the direction of energy flow create local pockets of potential energy and complexity (decreases in entropy) that serve to accelerate the degradation of energy within the system as a whole. Life exists for this reason. Living organisms are local concentrations of complexity and high energy density which process energy more thoroughly (meaning faster sometimes, more effectively sometimes, more efficiently sometimes) than do inanimate objects in the same open system. For example, rocks erode gradually on their own, but bacteria speed up the chemical breakdown of rocks considerably. A variety of local concentrations of living complexity provides more diverse opportunities for energy degradation; hence biological variation is also promoted. This can be seen in the greater biodiversity of the tropics where energy flux is greater, compared to arctic regions. Organic evolution is dictated by the need for ever more effective energy degradation, though given the variation introduced by random events (like genetic mutation), local conditions, and historical contingencies, the precise nature of those changes (i.e. the resulting characteristics of any given organism) is highly unpredictable.

But the overall trend is clear—enhanced facilitation of energy dissipation. The logic that governs inorganic and organic evolution also applies to behavior. The behavior of individual organisms is geared toward utilizing energy most effectively, and this generally means achieving the most energetically favorable outcome through the least amount of effort (least action). Again, however, behavior may in specific instances lead to local increases in complexity, as when behavior is used to build machines (like a bicycle) which enable more work to be accomplished per unit of behavioral effort overall. In addition, the actual methods by which the ends of behavior are achieved are a function of such unpredictable and indeterminate contingencies and antecedents that the appearance of volition and free will are very strong, even if in fact they really are deterministic in origin.

The same logic extends further to group behavior, and from there to social interactions and the structure and function of societies. At every level of organization, from individual to group to state to nation to global community, collective behavior is driven by the imperative to consume and degrade energy most effectively, often coupled with the accumulation of more information. The appearance of "emergent" properties is nothing more than sudden, significant advances in the operation of the PLA, in accordance with the SLT and in a manner appropriate for a given level of organization.

The evolution of humans represented a unique event in the history of life on Earth, in that a species emerged which, consequent to the mastery of controlled fire and the development of language, not only accelerated the ability to effectively use and degrade energy, but elevated the value of acquired intragenerational information to the level of a commodity of survival value. Since information links to entropy through communication and information theory, this is consistent with the imperative of the Second Law.

1.3 Meeting the Challenge of the Evolutionary Imperative

From the perspective of human livability, the accelerated expression of the Second Law has been too much of a good thing. Excessive energy consumption, while helping to achieve a comfortable living for many, has begun to degrade the environment to an unsustainable degree, and led to economic and social inequities that threaten social stability. One way to avoid the runaway catastrophe that would appear to be a realization of the Second Law with human complicity, is for humans to switch from energy consumption to information processing as a valued goal. This would enable the human species and the world of which it is a part, to continue to function within the mandates of the Principle of Least Action and the Second Law (as they must) but in a more benign and beneficial manner, since there are no inherent effects of information processing on resource depletion or on environmental pollution.

Summary of Chapter 1

The trend throughout space-time since the onset of our universe at the big bang has been for matter overall to become less concentrated and for energy to become more dispersed. This follows directly from the Second Law of Thermodynamics (SLT), a law of nature never found to have been violated.

Yet seemingly contrary to this, concentrations of organized matter and energy have appeared, ranging from spiral galaxies to ocean gyres, and all the forms of life on Earth. As matter and energy have become more highly organized at the local level, the information required to specify and regulate the components of the system has become greater, reducing rather than increasing the entropic state of the system.

The usual explanation for these apparent exceptions to the Second Law is that they constitute open systems that consume energy from their surroundings to decrease entropy locally at the expense of an increased entropy of the entire open system of which they are a part. *Why* the local decreases in entropy occur, however, is not something that the Second Law predicts or can explain. Thus, another perspective has recently been advanced—namely that entropy declines locally and in the short-term to allow a long-term increase in the efficiency and efficacy with which energy is dissipated. That energy dissipation *must* occur as effectively and efficiently as possible whenever gradients of potential energy are present is the consequence of an even older but less well-known natural law, the Principal of Least Action (PLA). Our contention is that the SLT and the PLA acting in concert provide an explanation for why evolution occurs and the direction that it generally takes.

Currently, the latter view, which we refer to as the **energy dissipation model of the evolutionary imperative**, has been expressed only theoretically and with few applications. The purpose of this book is to provide a systematic argument for the model and to apply it to a larger range of phenomena. We have examined the evidence contrary to and in support of the model, and found it to be generally applicable with strong explanatory power. At the end of the book, we extend the application of the model into the future by elaborating on the developing trends that are threatening the existence of our biosphere. Then we close with a proposal for increasing the likelihood we will be here long enough to receive the gratitude of our grandchildren for laying the groundwork for their survival.

2

Underlying and Universal Laws Governing Matter and Energy

Our universe is thought to have originated in a massive release of energy at a single point in space and time, known colloquially as the "big bang," about 14 billion years ago.[1] Ever since that unique and transformational event in space and time—what scientists call a singularity—the universe has been in a state of disequilibrium, meaning that matter and energy have been distributed unevenly. This is because the fabric of the universe unfolded from the big bang in a slightly less than perfectly symmetrical manner, probably because of minute quantum fluctuations at the outset. Over time, the slight unevenness in the distribution of matter and energy has set up gradients that provide an impetus for the concentration of matter and the flow of energy. For example, higher concentrations of matter exert a gravitational pull on less densely concentrated matter nearby; and heat flows from a warm body in which it is concentrated into the cooler surroundings where heat is more dispersed. These

rearrangements and flows occur in accordance with fixed laws of nature that, to the best of our knowledge, have applied everywhere and for all times throughout our universe.

Underlying all the laws of nature that govern the physical world are those that specify the way in which matter and energy can be transformed. While energy can exist in many forms, and can be converted from one form to another, it can neither be created nor destroyed.[2] From Einstein's Law of Special Relativity, however, we know also that matter and energy are interconvertible, through the well-known relationship, $E = mc^2$ (energy equals mass times the speed of light squared).[3]

Energy performs work by exerting a force. All the natural forces that bring about the rearrangement of matter and redistribution of energy can be traced to just a few that are considered "fundamental," in the sense that they appear to be empirical facts of nature with no other ultimate explanation or derivation. These forces account for every change that takes place in the universe. Thus, they represent the fundamental source of evolutionary change in both the abiotic and biotic world. However, not every change that could occur in every conceivable direction, actually does occur. This is because the forces of nature are confined to a particular pathway in a specific direction by a set of invariant physical constraints. Two of these—the Second Law of Thermodynamics and the Principle of Least Action—are our principle concern, because they provide, in our view, an explanation for not just how, but why evolution occurs.

2.1 Forces of Nature

Nothing about our world is static. The expressions of energy dissipation are manifest over all time scales. Every second, hearts are beating, plants are growing, and clouds are floating by. Over the course of a year, seasons change, houses are built or burned down, and 6900 cubic kilometers of water flow from the Amazon River into the Atlantic Ocean. Over decades, cities expand, governments rise and fall, and fortunes are made and lost in the manufacturing, service, and entertainment economies of the world. Over geological spans of time, mountains rise up and are eroded back to plains, oceans ebb and flow, the Earth's axis tilts first one way then another, and the Sun grows ever so slightly brighter on a course that will eventually cause it to engulf the inner planets of the solar system. And every one of these changes can be traced to a number of mathematically precise principles that can be counted on the fingers of a single hand.

Physicists today recognize four fundamental interactions that account for all the ultimate forces of nature.[4] The first is the "**strong**" nuclear force that holds the protons and neutrons of an atom together. The second is the "**weak**" interaction among subatomic particles that appears as radioactive decay and governs the emission of subatomic particles like electrons, positrons, and neutrinos. These two forces account for the way an atom's nucleus is held together and what happens when either fusion or fission of the nucleus occurs. They operate over extremely short distances, well below the scale of human perception. The other two fundamental forces of nature, by contrast, operate over an infinite distance and affect the world of objects and energy as we perceive them. One of these forces is **electromagnetism**, which governs the behavior of charged particles and appears as energy in the form of light, radiant heat, electricity, and magnetic attraction or repulsion. The final and weakest of the forces of nature is **gravity**. Though much weaker than the electromagnetic force, gravity operates over cosmic distances because all the positive and negative charges of large bodies (other than magnets) neutralize one another.[5]

The forces of nature collectively give rise to the concentrations of matter and energy that generate the gradients that lead to all the dynamic features of the universe. Atoms under the force of **gravity** are crushed together in stars, consuming mass and releasing energy. When hydrogen atoms fuse to form helium in the core of the sun by forcing four protons to bind together through the **strong force** as two protons and two neutrons, the **weak interaction** caused by the transition of protons to neutrons releases **electromagnetic radiation** in the form of light and radiant heat that warms the planet and drives photosynthesis fundamental to most of the Earth's living ecosystems.

Most of what we experience as change in the everyday world amounts to converting energy from one form to another, or redistributing matter through the expenditure of energy. Potential energy is converted to kinetic energy when a stone rolls down a hill. Chemical energy is converted to electromagnetic energy when fire emits radiant heat and light. Typing this page, for instance, involves several forms of energy. The kinetic energy of a rotating wire through a magnetic field generates an electromotive force carried by electricity, which heats a burner on the top of a stove that cooks the food that we consume so we can move the muscles of the fingers that type out a message that conveys information. All these conversions are possible because matter, energy, and information are related. While the forces of nature provide the impetus for their interconversions, however, not all the possible trajectories they could take in changing from one form, location, or state to another are possible.

This is because of additional laws of nature that determine the pathway and direction that changes in the natural world can take.

2.2 Laws of Thermodynamics

The subject of thermodynamics literally deals with the way the thermal properties of a system can change. The science of thermodynamics matured historically during the industrial revolution when heat was the major form of energy used to power machines. However, in reality the subject of thermodynamics extends to all forms of energy and encompasses the way in which both matter and energy behave when their state is altered in any way.[6] There are several basic laws of thermodynamics, of which the first and second are most central to our thesis.

The **First Law of Thermodynamics**, also known as the Law of the Conservation of Energy, states that the total amount of energy in the universe is constant. It follows that matter and energy, while interconvertible, can neither be created nor destroyed in their totality. It also follows that, since work consumes energy, the amount of internal energy that is retained by a system is equal to the energy it takes in from its surroundings, minus the work it performs.[7]

The **Second Law of Thermodynamics (SLT)** gives direction to all the changes in the flow of energy or the dispersal of matter that occur in any dynamic process. For our purposes, a useful way of stating the SLT is as follows:

Whenever energy is transformed, not all of it can be used for useful work, because some of it is dissipated as heat and the process of increasing the random arrangement of matter.[8] Because the amount of energy available for useful work is reduced any time energy is transformed spontaneously, the SLT is often viewed as a requirement that energy be "degraded" whenever it does work or changes its form spontaneously.[9]

The term "spontaneous" has a technical meaning here that differs somewhat from our everyday use of the word. In thermodynamics, a spontaneous change is one that goes from a higher initial to a lower final state of potential energy. Human bodies do not spontaneously combust (at least not very often

[10]), and a stack of firewood doesn't suddenly catch on fire without being ignited, even though the potential chemical energy of an organic body or firewood prior to burning is much higher than the potential energy left in all the carbon dioxide, water vapor, and charred remains following the conflagration. In both cases, energy from a lit match is needed to supply the "activation" energy needed to get the fire started. Once underway, however, matter in both cases will be transformed from complex organic molecules with high potential chemical energy to water, carbon dioxide, and residual carbon with a much lower total potential energy. By the definition above, these changes were spontaneous, even though they didn't spring forth without a small input of energy to get them started.

Often, the focus of the SLT is on the process of increasing the random arrangement of matter. The term for the degree of randomness, or disorder, of matter in a system is **entropy**.[11] When disorder increases spontaneously, energy is released in the form of heat and the dispersal of matter. It follows that for entropy to be reduced, or for order to be increased, energy must be applied or added into the system. Thus, the construction of anything that is more highly ordered (less random) than its surroundings—be it a structure, machine, or living organism—requires an input of energy. Dissolution of the structure, machine, or organism, on the other hand, occurs eventually of its own accord.

To summarize, the SLT says that whenever matter is moved, or energy is transformed from a higher to a lower potential, disorder (entropy) is increased and energy is released, which may or may not be used to do work. "Potential" can also be thought of as concentration of either matter or energy. The difference between a higher and a lower potential defines a gradient, so the SLT says that matter diffuses and energy flows always from regions of higher to lower concentration, when they do so spontaneously. The energy left over at the end of the process is degraded, because less of it remains for doing useful work than was present before the transition began. The gradient itself is thereby reduced as well.

Now let's consider some examples that illustrate the operation of the SLT.

1. Objects fall and roll downhill but never rise or roll uphill spontaneously. Higher potential energy due to elevation is converted to kinetic energy as objects fall down the gradient of gravitational attraction.

2. A sugar cube dropped into a cup of coffee dissolves and disperses throughout the volume of the cup. This occurs spontaneously as the molecules of sugar move from a compact, highly-ordered (low entropy) aggregate down their concentration gradient to more randomly dispersed individual molecules with higher entropy. In this case, very little energy is released; most of the change goes into increasing the entropy of the sugar molecules.

3. Once inside our muscles, sugar provides a high level of potential energy, which is converted spontaneously (with the help of enzymes that lower activation energies) to the kinetic energy of muscle movements by undergoing metabolic reactions that transform the higher chemical potential energy of glucose to the lower potential energy in the chemical bonds of carbon dioxide and water. The chemical energy from the glucose is degraded by its conversion into both useful (muscle movement) and non-useful (heat) energy, and the dispersal of its metabolic end products into the surrounding air, increasing their entropy.

4. A hot bowl of soup will cool to room temperature spontaneously. A cold bowl of soup, however, will not rise above the temperature of its surroundings of its own accord, even though there is plenty of heat energy in the combined air molecules surrounding the soup to heat it if somehow that heat could be gathered in from its dispersed location and concentrated into the soup bowl. The SLT allows soup to cool spontaneously because heat energy is dispersed into the larger volume of the surroundings. The SLT does not allow soup to draw energy from a more dispersed to a more concentrated location because that would entail a reduction in entropy.

5. Construction of a building requires a great input of energy, because so much material has to be arranged from highly dispersed origins into a compact and geometrically rigid structure. Apart from the lower entropy of the highly ordered structure, its elevation above the starting point of its component building materials gives it added potential energy. When it falls down, the potential energy inherent in its orderly structure and elevation is converted into the kinetic energy of its crumbling components, which end up in a much more disordered state than they occupied when the building was standing. If initiated by a few well-placed charges of dynamite, the building will collapse spontaneously; but the individual bricks and girders will never assemble themselves spontaneously into an intact building.

6. Anyone who has ever lived with a teenager knows that their room becomes disordered with little effort, while much more energy is required to keep it tidy. Though purely anecdotal and decidedly unscientific, perhaps this image conveys the extent to which the SLT governs our everyday reality more immediately than the technical examples given above.

2.3 Principle of Least Action

If you live on the fifth floor of an apartment building and drop a ball from your outdoor balcony, it will fall in a straight line toward the center of the Earth due to the force of gravity. If you let it roll down the inside stairwell of your building, it will bounce down the stairs along the steepest path it can take within the spiral confines of your stairwell. If you drop it at the top of a long hill, it will roll to the bottom along the steepest path it can find around the boulders, bumps, and trees along the way. Have you ever wondered why this is so? Couldn't the ball in principle take a curvilinear path from your balcony to the ground, or zigzag from one side of the hill to another? Intuitively you know it wouldn't, because you know that the shortest distance between two points is a straight line, and the quickest way to get through a series of obstacles is to move in as straight a line as possible from one point of resistance to another. Something inside your mind just "knows" that objects under the influence of a force like gravity will move spontaneously along the shortest path in the least amount of time possible.

In the world of quantum physics, what we know to be true is not at all inevitable—there is actually a virtual infinity of different paths the ball could take in getting from the top to the bottom of the hill. What we "know" will actually happen is based, not on our understanding of theoretical physics, but on our everyday experience. For the scientists of the 18th and 19th centuries, however, intuition was not a satisfactory explanation. Thus it was that over a period of more than a century, contributions from a number of mathematicians gradually lead to a rigorous understanding of what we now know as the **Principle of Least Action (PLA)**.[12] In its barest and simplest form, the PLA states that the shortest distance between two points is a straight line. If travel occurs at the same speed regardless of the path taken, the path between two points that is the shortest in distance also takes the least amount of time to cover. Thus, the PLA requires the displacement of matter or the flow of energy to occur over either the shortest distance or the least time possible, depending on whether we measure the pathway for those changes in terms of distance or time.

While making intuitive sense to us because it describes the world as we experience it so well, the PLA is in fact a profound concept with nuances and complex implications that can take many forms. However, they all have in common the requirement that any dynamic "action" be minimized. In the parlance of modern physics, the term "action" refers to the path taken by matter or energy in getting from one point in space-time to another. Thus, we use one formulation of the PLA when focusing on the energy expended or consumed in moving mass over a given distance, and another version of the PLA when describing the amount of energy transformed over the time required for the action to occur.

The first case was described by the renowned Swiss mathematician, Leonhard Euler (1717-1783), who defined "action" as the integral of momentum over distance.[13] In Euler's version, a given amount of energy will be expended over the shortest distance possible to get an object from one point to another. In our example above, the ball will fall from the balcony to the ground in the straightest line possible that the prevailing wind currents and any obstacles in the way will allow. At times when using the PLA to explain phenomena of this nature, we will say that the PLA results in the expenditure of energy in the most efficient way possible.

The French-Italian mathematician, Joseph-Louis Lagrange (1736-1813) worked with Euler to produce a version of the principle in which the "action" is the integral of the change from potential to kinetic energy over time.[14] From this perspective, the release of potential energy per unit of time is maximized to the extent allowed by the system. Stated another way, a system will transform the most energy it can within the time available. The stone will roll as far down the hill as it can, transforming the potential energy of height into the kinetic energy of motion, as fast as it can within the limits allowed by the force of gravity and the resistance it encounters along the way. We will describe this as energy transformation that is maximally complete.

Some systems use energy more efficiently, while others consume or conserve it more completely; and the two are not always mutually compatible. Tradeoffs are sometimes necessary. The Law of Least Action favors both, with emphasis on whichever facet of energy usage is required by the system in question. When referring to energy transformations that are either maximally efficient or maximally complete, or both, we will simply use the term "effective."

2.4 Complimentary Nature of the SLT and PLA

While the PLA by itself would appear to do a good job of describing the dynamics of everyday life, in one important way it does not. From a mathematical point of view the PLA is indifferent to direction. A ball could leap up from the ground to your fifth story balcony, or bounce up the stairwell, or climb back up the hill without violating the PLA, so long as it traced the minimal pathway possible from one point to another. Yet we know from experience that balls don't *spontaneously* leap off the ground, climb stairs, or roll uphill, any more than molecules of carbon dioxide and water reassemble with residual charcoal back into firewood, or that sugar molecules in solution reassemble themselves into a sugar cube. Another principle must be at play to harmonize the laws of motion and energy expenditure with the way we know the world actually works – one that specifies the *direction* that mass will move in or that energy will flow in when they do so spontaneously (without the input of additional energy). As indicated in section 2.2 that additional principle is the SLT.

On the other hand, while the SLT compels the direction of spontaneous changes, it says nothing about the path those changes take. Thus, a "spontaneous" fire still requires a lighted match to get it started; a slight input of energy is needed to activate the transition that will ultimately lead the overall system to a lower state of potential energy and a higher level of entropy. And while the SLT demands that a stone roll down the hill instead of up it, the pathway actually taken is not influenced at all by the SLT; that is determined according to the PLA as the steepest and quickest way down that the terrain allows.

We see, then, that change in the natural world is channeled by these two overarching principles working in concert—the PLA defining the pathway and the SLT mandating direction. If the change is spontaneous, energy will be released, work can be done, and entropy will increase. If the change occurs in the other direction, energy has to be supplied from the environment. The path taken in both directions will nevertheless be the most effective that physical constraints allow.

Our thesis is that the combination of these two profound principles of physical science account for all the organization that we see in the universe, and most significantly all of the evolutionary changes that have brought them about. Together they constitute what we refer to as the energy dissipation model of the evolutionary imperative.

Summary of Chapter 2

Our universe came into existence, according to the best prevailing model, about 14 billion years ago, in an energetic outburst of unimaginable power. At the very onset of this singular event, there was no mass, and all the forces in the universe were unified. Within tiny fractions of a microsecond, those forces segregated out into, first, gravity, then the strong and weak interactions that govern atomic nuclei, and finally electromagnetism. Only after the forces had separated were some of them converted into mass.

From the outset, the distribution of matter in the expanding cosmos was not perfectly symmetrical. Where subatomic particles were relatively more concentrated, atomic nuclei (protons and neutrons bound together by the strong force) were generated, and began to accumulate electrons through electromagnetic forces, forming atoms. The first atoms gave rise to all the matter of the early universe. Under the force of gravity, slightly more matter collected around granular nodes of concentration, leading ultimately to particles of dust that concentrated further until stars were born, releasing energy as heat, light, and other forms of radiation. In time, the universe was populated by innumerable pockets of concentrated mass and energy, serving as gradients that have propelled all the dynamic properties the world has ever seen, continuing to the present. Evolution is the term we give to the historical trajectory of the changes that reflect those dynamic features. So long as gradients of matter and energy exist in the universe, the universe will be in a perpetual state of change—and evolution, therefore, will be ongoing and imperative.

But the physical world is not capricious. When change occurs, whether through the motion of mass or the flow of energy, it does so within the constraints of a number of physical laws or principles which appear to operate throughout the universe and for which no exceptions are known. In the 18th and 19th centuries, most of these laws and principles came to be stated with mathematical precision. As an explanation for the evolutionary imperative, we find one of the principles and one of the laws to be particularly relevant.

The Second Law of Thermodynamics (SLT) is the first of these. It states that any time a concentration gradient of matter or energy gives rise to the movement of mass or the release of energy, it does so spontaneously in the direction that lessens the gradient. Matter diffuses and energy flows from a region of greater to a region of lesser concentration. In so doing, matter becomes more randomly distributed, and energy becomes more dispersed, resulting in a net increase in the entropy of the system and its surroundings. Change can occur in the opposite direction, but only if energy is supplied to the system from its surroundings.

The Principal of Least Action (PLA) is the other overarching rule for any and every dynamic change. It states that any change generated from a gradient of either mass or energy will discharge the gradient by moving mass or releasing energy along the path of least resistance, over the shortest distance, in the least amount of time, to the fullest extent possible. Matter will thus be reallocated, or energy transformed, in the most efficient and complete, or overall effective, way that the system will allow.

The PLA specifies the pathway, and the SLT defines the direction, for all dynamic events. These two rules that govern the way the physical world works, act in tandem to channel the nature of all evolutionary changes at every scale of organization.

References and Notes

[1] Kaufmann & Comins, 1996

[2] Landis et al., 2012

[3] Editors, 2012

[4] Trefil & Hazen, 2001, Ch. 13

[5] According to the prevailing view in cosmology, all the forces were united at the onset of the big bang. Gravity separated from the other three first, then the strong force separated from the remaining two "electroweak" forces. Finally, the weak and electromagnetic forces came apart. Only after the fundamental forces had separated from one another, were quarks able to begin to stick together to form matter. A satisfactory union of all the forces except for gravity has been accounted for with mathematical precision in the "Standard Model." Gravity has yet to be adequately incorporated into a single "theory of everything," though an attempt to do so has been suggested in the "Unitary Method" (Bern, Dixon & Kosower, 2012).

[6] Drake, 2012

[7] Daniels & Alberty, 1961. If no change in pressure or volume occurs, any change in internal energy is given by $\Delta U = Q - W$, where ΔU is the change in internal energy, Q is the amount of energy absorbed, and W is the amount of work done (Drake, 2012).

[8] While there are many ways to state the Second Law, depending on the particular aspect of a system that is the focus of attention, a simple and straightforward mathematical expression is given by the free energy equation, $\Delta F = \Delta U - T\Delta S$, where F is the free energy, or the energy available for doing work ("useful" energy) at a constant temperature, U is the total internal energy of the system, T is temperature, and S is entropy, or the degree of disorder in the system (Drake, 2012).

[9] Work is the effect of energy on material changes in the dissipative structure. Energy that is dissipated merely as heat or an increase in entropy (disorder) without any other physical effect on a system is said to be "non-useful" work; while energy that exerts a force on or alters the state of matter is defined as "useful" work. Energy is dissipated when work by-products are more disordered than the original substrate, e.g., gravel and silt in a river compared to the bedrock from which they were eroded, or CO_2 and ashes from a fire compared to the original wood. Thus the effectiveness of energy dissipation is a net sum of the effectiveness of "useful" work and the dissipation of energy as heat.

[10] Arnold, 1995a

[11] Entropy is a statistical concept. Technically, it is a measure of the extent to which all the elements of a system can exist in different states. Liquid water has more entropy than ice, because the individual water molecules in the liquid state can potentially occupy any space within their container, while in ice, the molecules of water have to be arranged in a specific geometric configuration with fewer degrees of freedom (Trefil & Hazen, 2001)

[12] The earliest formal description of the Principle of Least Action is generally attributed to Pierre-Louis Moreau de Maupertuis (1698-1759), the French mathematician and biologist recruited by Frederick the Great to head the Berlin Academy of Sciences. Contemporary mathematicians prefer to call it the Principle of Stationary Action, to convey the concept that the actual path that nature takes in moving matter or transforming energy coincides with the mathematically optimal path, hence the actual trajectory is "stationary" relative to the optimal path (Taylor, 2003).

[13] Euler, 1744. In Euler's formulation, action (S) is minimized by the PLA when $S = \int mv\ ds$, where mv is mass times velocity (momentum), integrated over distance. Since momentum represents the movement of mass over distance (work), this amounts to saying that a given amount of work will be performed in a way that consumes the

least amount of energy necessary. Stated another way, it says that the amount of work performed will be maximized per unit of energy consumed. We describe this as energy transformation that is maximally efficient.

[14] Taylor, 2003. A simplified version of the Euler-Lagrange equation states that action (S) is minimized by the PLA when $S = \int (K\text{-}P)\, dt$, where K-P is the difference between kinetic and potential energy, integrated over time. This means that energy is transformed to the maximum extent possible per unit of time, which we describe as energy conversion that is maximally complete.

3

Change in the Physical World

According to the standard model of cosmology currently embraced by most scientists, change began very early in the history of the universe. The first evolutionary [1] event for which a logical, objective argument can be made is the separation of gravity from the otherwise unified forces of nature, and models suggest that this occurred about one ten-trillionth of a yoctosecond (10^{-43} seconds) after the singularity at which point time began and space was created.

Gravity separated from the other fundamental forces when all of space occupied a volume less than the size of a proton. Then the strong nuclear force "froze out" from the electroweak force as the explosive expansion of the universe caused it to cool dramatically. By one yoctosecond (10^{-24} seconds), expansion had slowed, and subatomic particles were beginning to form the precursors of matter out of what had been, for want of a better term, pure energy. By this time, space had grown to a volume of about 100 km in diameter. About a nanosecond (10^{-9} seconds) into the life span of the

universe, the weak nuclear interaction separated from electromagnetic energy, completing the quartet of the fundamental forces of nature.[2]

In these earliest of events, where relativistic phenomena predominated over physical laws as we experience them on a human scale, the laws of thermodynamics and principles like that of Least Action are difficult to discern. In time, though, the chaos at the creation of the earliest slivers of space and time gave way to a progression of events that brought increasing complexity, diversity, and degrees of order to the evolving universe—a complexity and order attributable to crystallization of the laws of thermodynamics and the Principle of Least Action.

3.1 Atoms

By the time the universe had existed for about one second, it had expanded to a space of several trillion km (perhaps a few light years) in diameter, and atomic nuclei began to form. By the end of the nucleosynthesis era—about 200 seconds into the life span of the universe—an excess of protons over neutrons had accumulated. For the next two or three hundred thousand years, the universe consisted of an opaque soup of subatomic particles too dense for photons to travel unobstructed for any distance. At about 300,000 years into its existence, the universe had cooled enough for atomic nuclei to start capturing electrons, thus giving rise to the first atoms. As they did so, the opaque era ended, and electromagnetic radiation could begin to be propagated through the full extent of the universe, which was now roughly 100 million light-years across. Then, and only then, could we have begun to see the emergence of a world that we could have vaguely recognized.

As they began to capture electrons toward the end of the opaque era, the vast majority of resulting atoms were hydrogen from the left-over protons, and helium (two protons and two neutrons) in a ratio of about nine hydrogens to every helium. A very small fraction of lithium (3 protons and 4 neutrons) also formed. These three atoms constituted the entire universe of matter and made up the first generation of stars, before any of the other 95 naturally occurring elements had evolved at that time.

3.2 Stars

The fabric of the universe is a mystery. What we know is that mass has been distributed unevenly through space since its earliest inception, perhaps due to quantum fluctuations at the beginning of time. Because of that, the force of gravity, which is directly proportional to mass, has been slightly stronger where mass has been more highly concentrated. Thus, those small inhomogeneities in the distribution of mass (or "gravity wells" in Einstein's formulation) led to a slightly greater concentration of atoms in one region of space than in another. Since the mass associated with a greater concentration of atoms exerted an ever-stronger gravitational pull on less concentrated atoms in the vicinity, the aggregation of atoms grew. This growing mass of atoms, mostly hydrogen, grew larger and more massive, exerting increasing gravitational pressure toward the center of the mass until the pressure was great enough to force protons of hydrogen to fuse together. When the protons from four hydrogen nuclei fused into

a single helium nucleus of two protons and two neutrons, the total mass of the helium nucleus was slightly less than the mass of the four hydrogen protons. Since mass can be converted into energy but not destroyed, the difference was given off in the form of radiation (neutrinos, positrons, and photons). In other words, light and other forms of solar radiation began to be emitted, as hydrogen fused into helium, and a star was born.

This process occurred throughout the universe, wherever atoms aggregated in greater concentration at one point than another. These initial aggregates appear to have been generally more massive than the newborn stars of today, but they were simpler in composition because the only elements available were hydrogen, some helium, and a tiny fraction of lithium. The huge size of the stars, however, meant that the gravitational pressures at their core would have been overwhelming enough to consume all the hydrogen relatively rapidly. Once all the hydrogen was exhausted by fusion into helium, the helium atoms themselves would start fusing into larger nuclei, which in turn could form even larger combinations, such as the nuclei for carbon, nitrogen, oxygen, sodium, silicon, chlorine (with 6, 7, 8, 11, 14, and 17 protons, respectively), and all the other elements up to the size of iron (26 protons). The formation of all heavier elements was driven by gravity as star matter compressed upon itself, down the steepest potential energy gradients possible in accordance with the PLA. These atoms would eventually be flared off from the stars, or violently expelled from their cores into interstellar space when the stars exploded as supernovae, then distributed by solar winds and diffusion throughout the cosmos as dictated by the SLT. The power of supernova explosions is great enough to form even larger elements than iron, despite the fact that energy is consumed rather than expelled in their formation. All the elements, and especially the larger ones, could serve as grains of condensation for new aggregates of matter that would coalesce into new stars, now with a richer chemical composition. Through repetitive cycles of star formation, destruction, and reaggregation, the universe became enriched by the addition of all the elements that exist today, and are still being created. [4]

3.2.1 Large Scale Structure of the Universe

Peering into the depths of a starry sky at night gives the impression that the stars are scattered randomly across the heavens. But they are not. Because the distribution of mass across the universe is uneven,[5] and because the force of gravity is proportional to mass, matter is drawn into clumps where the pull of gravity is strongest. The appearance of random scatter among the stars is an artifact of the small sample of the galaxy that can be seen with the naked eye. Save for a few faint blurs of light, every easily visible star in the night sky is part of the Milky Way Galaxy to which the sun and its solar system of planets belong. The Milky Way, like all galaxies, is actually a non-random concentration of stars (estimated at 100 to 400 billion). Between the galaxies lie vast distances largely devoid of other stars. A view from outside the Milky Way would therefore reveal that the stars are scattered through space in highly non-random concentrations, and that these concentrations (galaxies) themselves form non-random clusters, filaments, and aggregates of light, heat, and atomic nurseries, presumably reflecting the large scale, uneven distribution of mass that came into being during the earliest gestation of the universe (Fig. 3.1).

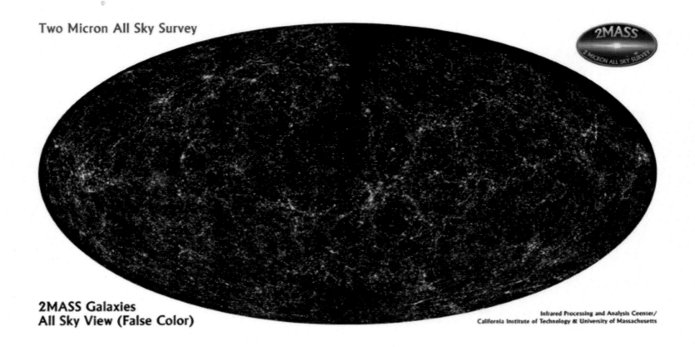

Fig. 3.1. Large-scale structure of the universe. This panoramic view of the sky plots over a million galaxies beyond the Milky Way. Bluer galaxies are younger and nearer; red ones are older and further away. Note non-homogenous distribution of clusters and strands. *Two Micron All Sky Survey/Caltech/University of Massachusetts*

At the simplest level of analysis, we can view the growing complexity of the large-scale structure of the universe as a manifestation of the force of gravity, acting through the constraint of the Principle of Least Action (PLA). Once gravity wells are formed, they pull matter toward them. Galaxies are thus arrayed along contours dictated by the uneven distribution of mass in the universe. The fact that the stars align according to these contours is simply a reflection of the fact that matter will move along a force field gradient in the straightest possible line, as dictated by the PLA.

3.2.2 Galaxies

From a larger, more encompassing vantage point, the galaxies themselves have more complexity than the appearance of random scatter in the nighttime sky would suggest. First, they fall into a number of discrete shapes and arrays—some being elliptical, others forming spirals of two or more strands, while others appear disorganized. Secondly, they move; stars within a galaxy tend to rotate around their galactic centers, and the galaxies themselves are moving in relation to one another.

There is growing evidence to suggest that the shape of a galaxy is an indication of its age.[6] The oldest galaxies consisting of the most ancient stars (so identified by the very small proportion of elements larger than lithium within them) tend to be smaller and less organized or elliptical. Spiral galaxies, like the Milky Way, appear to reflect a more evolved state in which smaller galaxies have been drawn in by the gravitational pull of larger galaxies.[5] Within certain regions of spiral galaxies, star formation is particularly active, so new stars with higher metal content are being formed at a higher rate. A crude evolutionary sequence among shapes of galaxies can thus be envisioned (Fig. 3.2), with spiral galaxies like the Milky Way representing a more recent, more complex, and more highly evolved galactic structure.

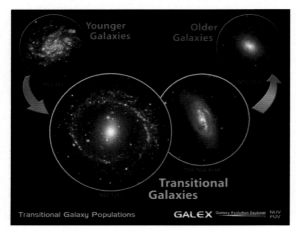

Fig. 3.2. Galactic evolution. In their youth, galaxies are smaller, denser, unorganized arrays of stars with much star-forming (blue emitting) activity, like NGC 300. In time, they are thought to undergo a series of collisions and interactions that lead to increasingly complex, often spiral, structures, like NGC 1291 and 4569. Spherical galaxies, like NGC1316, represent more mature stages, with diminished star-forming activity. *NASA/JPL-Caltech/CTIO/Las Campañas/Palomar*

The second fact of galactic dynamics is rotation – not only of galaxies as a whole but of the stars within them. As we shall see, the planets that form around the stars move in orbits about their solar centers of rotation, and spin on axes of their own. Such motion must be deeply rooted in the physical laws of nature. Indeed, the principle of angular momentum dictates that matter move in a circular motion when it accelerates toward a center of attraction. This can be seen as a corollary of the PLA, because a curvilinear path toward the center of attraction represents the shortest, steepest path possible down the gravity gradient that pulls matter in from all possible directions. At the level of common experience, as a bathtub empties, the water swirls around the opening into which it drains. From this highly local example to the galactic scale, the same principle applies. In the aggregate, curvilinear paths of movement around a center of attraction give rise to rotation. Thus, where once there was movement that appeared to be random, under the influence of gravity gradients and the constraint of the PLA, matter is led to move in a more complex and ordered manner.

3.3 Planetary Bodies in Their Variant Forms

While the vast majority of matter in a protostellar cloud gathers gradually into a mass at the center of the spinning disk, forming its central star, some of it aggregates in the periphery. Small grains of dust clump together and grow progressively larger through sequential collisions of increasingly larger clumps of matter, until essentially all the smaller particles of dust, rock, and water within a particular radial trajectory around the central star are aggregated into a mass large enough to form a metallic globe surrounding its own gravity well. Thus a planet—a satellite of the central star—is born.

Those planets that form far enough from their central stars to be cold enough for gasses like nitrogen, methane, helium, and hydrogen to be liquefied, form gas giant planets, like Jupiter and Saturn. Planets close enough to their stars to be warm enough to vaporize the smaller gas molecules, and low enough in mass so that gravity is insufficient to hold them, turn into rocky planets, like Mercury, Venus, Earth, and Mars. Venus and Earth are massive enough to hold onto larger gas molecules, like N_2, O_2 and CO_2. Mars, with a much lower mass, barely sustains a very thin atmosphere of the largest of these gases, CO_2.

Our Solar System fits nicely into this idealized model of the size, nature, and distribution of planets around a central star: small, rocky planets nearer the Sun, gas giants further away. The ongoing discovery of exoplanets in other solar systems, however, is painting a more complicated picture. Given the fact that very large planets are easier to detect than smaller ones at distances light years away, the preponderance of exoplanets discovered to date are much larger than Earth. That doesn't mean there aren't as many or more smaller planets out there. Somewhat more surprising, though, is the discovery that many of the huge exoplanets actually circle their central stars at distances well within one astronomical unit (the distance from the Earth to the Sun); so many of the exoplanets, which must be gas giants, are circling much closer to their suns than our gas giants orbit ours. Modeling the structure of solar systems beyond our own is obviously a work in progress.

Minerals, like all facets of the natural world, evolve through time. There are currently close to 4,400 known mineral compounds making up the Earth, but that wasn't the case in the beginning; when the Solar System formed there were only about a dozen. The diversity of minerals increased over time through the successive action of several processes. During the early stages of planet formation, once the Sun ignited and started bathing its planetary satellites in pulses of heat, successive waves of melting and re-mixing brought the planet's core elements into new combinations, including the first iron-nickel alloys, sulfides, phosphides, many oxides and silicates. These chemical combinations represented more stable aggregates of free energy—consistent with the SLT—but also more variety and complexity. This diversity increased progressively through ongoing melting and re-mixing during the "black Earth" stage of planetary evolution that characterized the Hadean Eon (4.5 to 4.0 billion years ago). With plate tectonics generating cycles of eruption and subduction over the crust of the Earth, remelting and remixing the planet's inventory of minerals, the number of mineral compounds rose close to 2,000 by 2 billion years ago.[7] By that point,

photosynthetic organisms had begun to infuse the atmosphere with oxygen, a highly reactive element that brought on the "red Earth" phase of oxidized compounds, doubling the number of different minerals through the action of living organisms, as we will elaborate further in the next chapter.

3.4 Geophysical Change Through Time

The planet beneath our feet is restless. We're reminded of that by the catastrophic earthquakes that strike somewhere on Earth every couple of years, on average—with devastating consequences where populations are concentrated. Residents of California get a gentle reminder of this every few days. A volcano erupts with explosive force somewhere every few decades, blowing cubic acres of magma and earth into the sky, and covering the surrounding surface with lava that hardens into newborn substrate. Sometimes lava erupts from the ground and creates a mountain *de novo*. Mounts Fuji in Japan and Kilimanjaro in Tanzania were formed in this way, as was Mount Olympus on Mars, the largest known volcano in the Solar System.

Earthquakes and volcanoes are the sudden, perceptible expression of the fact that the crust of the Earth is in motion. Massive continental plates make up the crust (lithosphere), and float on a more plastic, partially molten layer (asthenosphere) beneath the crust,[8] something like the sections of a cracked egg shell, were the pieces of the shell able to bump and grind against one another as they slide about on the liquid white inside the egg. Some plates are pulling apart, letting magma from below erupt through the widening gap between the separating plates, forming new substrate as the original plates move in opposite directions. The North and South American Plates are pulling apart from the African and Eurasian Plates at about two centimeters per year in each direction along a ridge under the Atlantic Ocean, creating the mid-Atlantic ridge.[9] Volcanic eruptions near the northern extent of this ridge created Iceland. At other points, the plates are crashing into one another, causing both to crumple up, or forcing (subducting) one beneath the other. The Andes Mountains were formed by the crumpling collision of the South American plate and the oceanic Nazca Plate. The hot springs and geysers of Yellowstone National Park in Wyoming are produced by heat from the friction of the Juan de Fuca Plate as it slides beneath the overriding North American Plate. Wherever geysers are spewing, or volcanoes are erupting, heat from the Earth's interior is driving water or magma up through weak spots in the crust. The newest of the Galapagos Islands and the big island of Hawaii were formed entirely this way barely a million years ago, with Hawaii still expanding by this process on a daily basis. Whenever earthquakes occur, they are relieving stress along the fault lines, or cracks, in the crust. All these processes have in common the fact that heat from the Earth's interior is driving magma to the surface, or heating underground water reservoirs, or causing convective movements beneath the Earth's crust that move the continental plates around. Over geological time spans, all these dynamic geophysical processes rearrange the landscape—building up mountain ranges, creating islands in the oceans, and opening up gaps in the land that are filled by seas.

What is the source of the heat that drives this activity from the Earth's interior? Some of it is the reservoir of heat derived from the kinetic energy of meteors and planetessimals striking the Earth during and following its formation. Much if not most of it today comes from radioactive decay of heavy metals in the core. When radioactive materials decay to lower energy states spontaneously, in accordance with the SLT, they release energy from the weak interactions in the atomic nucleus. This energy manifests itself largely as heat, conducted by magma toward the surface. The energy gradient thus generated by two of the fundamental forces of nature (gravity, which concentrates mass at the center of the planet, and the weak interaction, which provides the energy released by nuclear decay), is resolved by moving heat from its source in the Earth's core to a cooler and more dispersed area at the planet's surface. Currents of convection following the paths of least resistance outward from the core according to the PLA, are energized by this flow of heat through the liquid mantle, and provide the motive force for geysers, volcanoes, earthquakes, and presumably all of the processes associated with movement of the tectonic plates. Thus, the land and oceans of the Earth are reconfigured as energy

flows down its concentration gradients, degraded as demanded by the SLT, along convection currents channeled by the PLA.

Fig. 3.3. Evolution of the planetary surface. Mountains are thrust up, perhaps from ancient sea beds indicated by multiple layers of sedimentary deposition (upper panel), by forces emanating from the planet's interior. Wind and water, driven to resolve pressure or gravity gradients, erode higher elevations toward the lowlands (middle panel), creating in time broad plains of infill (lower panel). *Credit: photography of Glacier National Park in Montana by Louis Irwin.*

3.5 Evolution of Planetary Topography

The geophysical history of the Earth suggests that these large-scale plate tectonic events occur in cycles, aggregating the land into a single massive continent, then splitting it into fragmented continental plates that wander their separate ways, through cycles that repeat on time scales of a half to a billion years. The mountains that are built up and the shorelines that are created by these cycles of continental aggregation and refragmentation are altered over much faster time scales, however. As soon as mountains are created, they begin to wear down. The land is alternately flooded, desiccated, or cut by flowing streams. A seashore is created as sea levels fall, then eroded as the sea reclaims the shore. Thus the surface of the Earth is in a constant state of change.

3.5.1 Mountains

The process of orogeny, or mountain building, uses the dissipation of energy from the planet's interior to raise the potential energy of the mountain itself. Overall, there is a net reduction of useful energy, as demanded by the SLT, but locally a new potential energy gradient is created by the elevation of the mountain. This gradient in time will ensure its demise.

Any natural force that has the capacity to move the soil, like wind or rain, will begin the process of eroding the mountain away. The potential energy intrinsic to the mountain's elevation provides a gradient that the downward movement of matter resolves. Snow and rain falling at upper elevations turns to bodies of water that flow downhill, cutting rivulets, then streams that carry soil, rocks, and boulders. Wind blows soil from the upper reaches of the mountain to an ultimate deposition at mostly lower elevations. Gradually, the mountain evolves from an elevated structure of high potential energy to an eroded landscape that, in the extreme, is worn down to a state of equilibrium with the surrounding countryside (Fig. 3.3). The compact mass of the mountain is thus transformed by the dispersal, or increased entropy, of its matter; the bedrock formations at the upper reaches of the mountain get broken into a multitude of boulders, then pebbles, then sand as matter makes its way downward toward the plain or sea where it will end up in a much more highly dispersed state. The potential energy of the mountain's height is converted through the kinetic energy of downward moving soil and rocks to lower final levels of potential energy. Mountains are built up through prodigious forces that depend on massive inputs of energy. They are then spontaneously degraded over time by the elements. The fact that mountain building is an energy consuming process, while the erosion of the mountain is an energy-releasing, spontaneous process, is a clear manifestation of the SLT.

3.5.2 Streams, rivers, and canyons

Water flows downhill spontaneously under the influence of gravity. The physical law underlying this truism is the SLT, since the water and whatever it carries move from a higher to a lower level of potential energy. In so doing, however, the water and its contents do not flow randomly – they take the path of least resistance (Fig. 3.4). Water flows downhill along the steepest, straightest path it can take, as demanded by the PLA, though obstacles in the way of that steepest straightest path ensure that the flow will trace a meandering path.

Fig. 3.4. Evolutionary action of water. Flowing downhill to resolve gravity gradients, water takes the path of least resistance, starting as a stream (upper left), and becoming a river capable of cutting canyons that channel the water more efficiently along the shortest path possible from higher to lower ground (upper right). At a few places on Earth, the crust is pulled apart, creating river valleys like that of the Rio Grande Rift in New Mexico (lower left). Though usually taking long periods of time to sculpt the surface of the planet, catastrophic outflows of water can sometimes create very large drainage systems when a natural dam, like the ice-aged glacier that contained prehistoric Lake Missoula, gives way, leaving the Columbia River Gorge as a remnant (lower right). *Photography of Rocky Mountain rivers by Louis Irwin (upper); Yuri Fialko / NSF (lower left); John Russell / USDA (lower right).*

Rivers form as streams converge into larger pathways of downward flow. River channels grow in size to the extent that they represent the path of least resistance regionally for an ever-growing volume of flowing water. As the volume of the river grows, it cuts the channel more clearly and deeply, all the while eroding the substrate and increasing the entropy of the soil and rock through which it passes. The progressive convergence of rivulets into streams, and streams into rivers, generates a complex, dendritic pattern which is far from random (Fig. 3.5). This is because the PLA is the antithesis of randomness, since only one of all theoretically possible pathways satisfy the requirement of "least action" (minimal deviation from the straightest path allowed by topography). It is this interaction between the tendency of the SLT to maximize randomness and the mandate of the PLA to minimize it, that accounts for the paradoxical increase in order and complexity as change occurs at a local level, all in accord with both the SLT and the PLA.

Canyons represent the ultimate dissipative force of a river. A great deal of work is done by the river as it carves out the optimal path for it to take down its potential energy gradient. While the work of eroding the canyon slows the dissipative process by diverting free energy to create the structured walls of the canyon, this process supports the dispersal process in the long run, in that the work leads to a more effective conduit (the canyon) for future energy dissipation.[10]

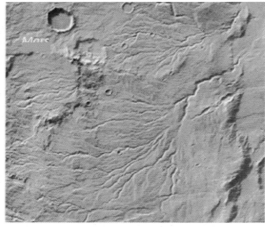

Fig. 3.5. Drainage basins. Water flows downhill, along the optimal paths available to it under local circumstances, whether in the Amazon basin or on Mars. *NASA (upper); NASA/JPL (lower)*

3.5.3 Ponds, lakes, and seas

Wherever water collects, from the smallest pond to the largest ocean, it does so at the lowest level of potential energy locally available to it. While the form of a pond or the shape of a shoreline may appear to trace a random path along the water's edge, that contour happens to form the perimeter of the basin that holds the water at its lowest point of potential energy. Since water is a liquid, every molecule of it moves by gravity to the lowest point it can occupy, resulting in a smooth and level surface at the same height as the interface between water and land around the perimeter of the basin that holds the body of water.

This tendency for water to seek its lowest possible elevation (minimal potential energy) within the basin that contains it can generate nicely patterned terraces when a series of containers are gathered in close proximity (Fig. 3.6). On a larger scale, the Great Lakes along the United States-Canadian border form an orderly sequence of large bodies of water, from Lake Superior at the highest elevation, through Lakes Michigan, Huron, and Erie, to Ontario at the lowest level. From there, the water continues its flow out the St. Lawrence Seaway to the Atlantic Ocean, at a lower elevation still.

The shapes of lakes and the shorelines around the seas and oceans of the Earth change over time. As the volume of water increases from inflow or decreases from evaporation, the size of the lake and the relief of the shoreline change. Or the water body's perimeter may be altered by other forces. The high level of earthquake activity along the Pacific shore of western South America frequently adjusts that shoreline by meters at a time. Darwin noted the elevation of one island by over three meters in a single earthquake, and observed other sudden instances of submergence and emergence of shore lines during his voyage on *The Beagle* in 1835.[11] Huge prehistoric bodies of water, like Lake Bonneville where Salt Lake City sits today, and Lake Missoula in what is now western Montana, disappeared in an instant, geologically speaking, when one edge of the basin holding them gave way. In the case of Lake Missoula, the resulting catastrophic outflow gouged channels many kilometers wide across eastern Washington. Even larger outflow channels can be seen on Mars, where huge bodies of (perhaps frozen) sequestered water have broken free periodically from their reservoirs to flow across vast areas of land, reconfiguring the landscape in the process.

Fig. 3.6. Order in nature. Gravity and the PLA combine to create complexly arrayed but ordered pools of water, as walls formed from calcium carbonate precipitating out of hot springs form a series of containers for water that spreads to its lowest possible level of potential energy within each container (upper panels). Another non-random structure is created when magma cools as it nears the surface (lower). Magma can contract vertically, but not horizontally except by fracturing along vertical lines. Columns of solidified basalt are then revealed when overlying earth is cut away by wind and water, revealing a highly ordered pattern of vertically fractured columnar basalt above horizontal layers of sedimentary gravel. *Photography of Yellowstone National Park, Wyoming, by Louis Irwin.*

3.5.4 Meadows, basins, and plains

As lakes fill in with silt and water levels drop, then disappear, meadows form. On planets like our own where life thrives, this process is accelerated by the growth of plants, which capture more soil and create it through their decay. The succession of lake to meadow to forest is especially easy to see in montane regions, where surrounding mountains provide ample mass for erosion and infill. As long as the mountain building process is active, of course, new lakes are created occasionally, at the start of a new sequence of succession.

On a larger scale, basins are created in the spaces between mountains when they arise. This typically occurs when the land pulls apart, as the crust of the Earth is fractured, then pulled in opposite directions by forces in the asthenosphere beneath the crust. Rift valleys provide the classic case of basin formation by this process. The Rio Grande Valley, from Colorado to the Big Bend of Texas is a good example in North America of where geological activity has created a broad basin between a series of mountainous ridges and elevated plains (Fig. 3.4, lower left). A mighty river and large tributaries once coursed through the land between the mountains and the plains to both the east and the west, as evidenced by the huge but now mostly dry arroyos on either side of the rather paltry river that flows through the valley today. As the climate dried and the surrounding land was eroded into the rift valley, it filled in to create the basin that it is now. Wherever mountains are created or the land is lifted up by the roiling forces beneath the Earth driven by the dissipation of heat energy from the interior, deep valleys are typically created, only to fill in eventually by erosion from the surrounding elevations, forming broad basins of essentially level land.

On a larger scale still, the great continental plains are created in the same way. The vast central plain of North America that stretches from northern Canada to Mexico is essentially infill from the older mountains and hills in the eastern part of the continent, and the younger mountain ranges in the west. The Ohio, Mississippi, and Missouri River valleys are the major remnants of an extensive riparian system that remodeled the topography of North America, filling in what was once a vast inland sea.

Whether meadow, basin, or plain, the successive waves of erosion and infill provide a striking example of how order can emerge from a process driven by increasing entropy. As matter is spontaneously redistributed from the heights of the terrain that surrounds a basin or plain, its potential energy is lowered and its contents are dispersed, in accordance with the SLT. In settling toward an equilibrated elevation, however, each particle is pulled by gravity to its lowest point as demanded by the PLA, causing the bulk material to spread out uniformly into level layers. Over time, the layers become superimposed on one another. The long periods that constitute geological time intervals are characterized by differences in atmospheric conditions—an extended period of volcanism, for example, would see more ash transported in the air around the globe—and by different source material, as the erosion of one type of geological formation gives way to another. These differing conditions give the layer of soil laid down during a particular period its unique character and coloration. We then are able to see the consequence of these processes when the long-buried layers of crust are uplifted (Fig. 3.3, top panel), or new rivers cut canyons through the layers of sediment laid down long ago (Fig. 3.7). What is now west-central North America was covered by a large inland sea during most of the Mesozoic Age (140 to 65 million years ago). Since the end of the Mesozoic, that portion of the continent has been filling in with successive layers of sediment from all directions. Over time, a series of orogenies, uplifts, canyon-cutting glaciers and rivers, wind sculptures, and modern-day road cuts have revealed the beautifully ordered sedimentary formations found throughout the Midwestern and western regions of the continent today (Fig. 3.7). Similar formations scattered among all the continents of the Earth, as well as on Mars—the only other rocky planet known to us to have been sculpted by water—give vivid testimony of how the SLT and the PLA work in concert to mold the ordered and increasingly complex features of an ever-evolving world, even as the entropy of the world as a whole increases.

Fig. 3.7. Sedimentary formations. Erosion reveals the sedimentary formations laid down by successive waves of infill into the great inland sea of the Mesozoic in what is now North America. *Photography in Theodore Roosevelt National Park by Laura Thomas, USA National Park Service*

3.6 Weather and Climate

The segments above should make clear that the physical world is in a constant state of evolution, driven by the dissipation of energy in its various fundamental forms, in the direction demanded by the SLT, along a path required by the PLA. Except for the first one or two seconds of the history of the universe, these physical changes have occurred over time intervals too long to be perceptible, were any sentient creature present to perceive them. Earthquakes and volcanic eruptions can be sudden, of course, but they depend on the buildup of forces that stretch across millennia, at a pace well below that of human notice. The one aspect of the physical world that does change at a rate perceptible in real time is the weather. And, as we've seen, weather plays a major role in the way the topography of the land is molded over time.

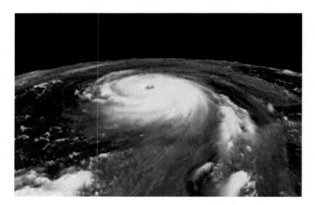

Fig. 3.8. Hurricane Katrina. A highly-organized mass of violently rotating wind and rain bears down on New Orleans on 28 August 2005. Hurricanes provide a particularly dramatic example of organized processes that resolve energy gradients (differential atmospheric pressures) in the physical environment. *NOAA.*

Weather refers to the dynamic activity of the atmosphere, in conjunction with the water cycle, in a local area. Wind, rain, drought, snow, ice, hail, storms, and the daily cycles of heating and cooling, are the changes in the physical world of which we are most immediately aware, because they change over time intervals of minutes to days rather than centuries to eons. Like all the other changes in the physical world, they trace to the same fundamental forces of nature, channeled as always by the SLT and PLA.

Weather is a property of the troposphere—the lowest level of that atmosphere extending to a height of about 8 kilometers (5 miles) at the poles and 17 kilometers (11 miles) at the equator. Within this strata, the clouds are concentrated and the air is densest, so this is where the wind is strongest, where the water evaporates from the surface then falls back to it as snow, rain, and hail, and where temperature fluctuations are greatest. However, gale-force winds (jet streams) blow at higher levels over longer sustained distances, with significant effects on the atmospheric dynamics below.

Weather dynamics are driven by solar radiation (produced by the weak interaction in the nuclear fusion process), but are modulated significantly by conditions on the planetary surface. As land or water are heated by radiation, the air over the surface warms and therefore expands. This reduces its density, and air with a lower density weighs less, exerting less pressure at the surface. This is what is meant by an atmospheric "low." When air temperature cools, the density of the air goes up, and pressure increases, generating an atmospheric "high." The difference in atmospheric pressure between adjacent highs and lows constitute the gradient that provides the motive force for movement of air, or wind. Morning and evening breezes reflect the gradients caused by warming air as the sun rises, and cooling air as the sun sets.

Stronger winds are generated by stronger gradients. Over the continental land masses, far from the thermal buffering effects of the ocean, temperatures rise much higher in the summer and drop much lower in the winter. In those parts of the globe where seasonal variations in temperature are significant, the transition from winter to summer is occasioned by the most extreme clashes between cold and warmer masses of air. This is why tornados are most likely in mid-continental land masses in the Spring. A different force is at work in the tropics, where prolonged periods of warmth cause extreme low-pressure cells to form over the ocean near the end of summer. As the warm air rises, it starts to rotate due to the Earth's rotation and the same conservation of angular momentum that accounts for the rotation of planets, solar systems, and galaxies. Only this rotation is more local and more rapid

by orders of magnitude, resulting in a hurricane (Fig. 3.8).

Precipitation occurs because the ability of air to hold water vapor depends on its temperature. Warm air holds more water, while cold air causes it to condense and fall from the sky as rain, hail, or snow. As water flows from wherever it falls or melts to the lowest elevation it can reach, it carries with it eroded soil from the land through which it passes. The energy from the sun is thus dissipated through the evaporation of water, the force of wind set in motion by differential gradients of atmospheric pressure, and the inevitable tendency of matter to scatter in an increasingly random manner when spontaneously dispersed by the forces of nature.

Climate is the summation of atmospheric activity and related environmental conditions, including solar radiation, temperature, humidity, precipitation, and wind averaged over a longer period of time. It strongly depends on latitude, proximity to water, and the topography of the land. Tropical climates are concentrated nearer the equator, where sunlight strikes the Earth at a more consistent, less variable angle. A greater range of climatic variation is seen at mid-latitudes where temperature oscillations with the seasons are greater. Seasonal variations, in turn, are caused by the tilt in the Earth's axis, which results in stronger, more direct sunlight when a hemisphere is tilted toward the sun in summer, and weaker, less direct sunlight when the tilt is away from the sun in winter. Because the Earth wobbles on its axis over long term cycles — every 40,000 years, the tilt reaches 24.5 degrees — the average temperature of the planet drifts up and down over these longer time intervals. At the present time, the Earth is enjoying a warm interlude between cycles of glaciation. Changes in the eccentricity of a planet's orbit (how close to the center of its orbit the central star that heats it is located) can affect global climates over longer periods of time. The Mesozoic, for instance was a period of nearly two hundred million years in which the Earth was generally warm, while the more recent 65 million years of the Cenozoic Era have seen considerably cooler climates overall.

In a later chapter, we will discuss how these normal oscillatory cycles in climate can be, and in fact now appear to be, in the process of being overridden by human activity, with potentially disastrous consequences.

3.7 Reflections on the Nature of Changes in the Physical World

The description above of changes in the physical world over time is intended as an argument for the fact that the material universe has been evolving since the earliest fractions of a second into its existence, under the primary forces of nature in a manner that increases entropy overall as the universe expands and cools according to the dictates of the SLT. In requiring that matter move and energy flow along the steepest, straightest paths possible, however, the PLA imposes a sense of order that frequently manifests itself as growing complexity – the symmetrical beauty of a shield volcano, the ordered layering of sedimentary rock, the organized violence of tornados and hurricanes, the meandering but maximally efficient path that water takes in flowing through canyons, being typical examples.

Over the course of their existence, planets can sometimes show directional changes. There is every reason to suppose that Venus and Mars were both once well-endowed with surface water, as the Earth still is.[14] Yet Venus in time became too hot to maintain liquid oceans, and Mars became too cold, with an atmosphere too thin, to retain liquid seas. The Earth itself has gone through cycles of cold and warmth—oscillating between snowball Earths and globally tropical climates.

The nature and extent of every change in the physical world builds upon the starting conditions from which those changes develop. The chemical elements created by an early generation of stars determines the atoms created in the next generation of stars. The canyon that begins to be carved by a river flood affects the course and breadth of the next flood. A strong sequential dependence can be seen in the development of galaxies (Fig. 3.2), the deposition and erosion of sediments (Fig. 3.7), and the shapes of mountains thrust up through pre-existing terrain. But all these changes, it should be noted, appear to be guided by stochastic processes only—by the operation and confluence of basic physical forces

operating with no direction or imperative apart from the dictates of their starting conditions.

On a dynamic planet like the Earth, repetitive processes occur over time. Some, like mountain building and erosion, span hundreds of millions of years. Others, like the buildup and advance of glaciers which carve changes in the landscape, occur over tens to hundreds of thousands of years. Still others, like hurricanes and tornadoes have life spans of days to minutes, respectively. They all recur, but each event is a new creation of nature, independent of any similar event that has gone before.

This is a critical point about evolution in the physical world that deserves emphasis. Each occurrence of a physical change is essentially an independent event, bearing no necessary relationship to similar events that have preceded it. Mountains come and go, but no two mountains are the same; the structure of a mountain chain arising in roughly the same region as an older chain many millions of years earlier (as in the case of the Rockies), bears no detailed relation to the former iteration of the same general phenomenon. While hurricanes and tornados are highly organized and regrettably recurrent, the microstructure of each one is totally unique – no preexisting storm dictating the detailed structure of any that follows. There is no perpetuation of information from one mountain, glacier, or hurricane to another. And even when directional changes in climate and topography have appeared over long periods of time, they have done so in response to strictly contemporary physical forces and trends. While they may retain a record of historical events, as the layers of sedimentary rock tell of ancient seas, that historical record has no influence on the way the future unfolds.

With the evolution of living organisms, that situation changed, as did the nature of evolution itself. That is the subject of the next chapter.

Summary of Chapter 3

Once gravity separated from the other fundamental forces of nature within a fraction of an instant at the beginning of time, evolution of the physical universe was underway. It continued with separation of the other three forces (the strong, the weak, the electromagnetic) and the emergence of the subatomic particles that would start forming protons, neutrons, and electrons. By 300,000 years into the life span of the universe, these were consolidating into the hydrogen, helium, and lithium that would constitute the first atoms.

As the infant universe expanded and cooled, clumps of slightly greater concentrations of atoms served as points of attraction for surrounding atoms. When gravitational collapse forced a large enough mass of hydrogen to start fusing into helium, the first stars were born, and the universe evolved into a billion points of light. Many stars became massive enough to forge the fusion of helium into larger atoms, like carbon, oxygen, and silicon. When those stars exploded at the end of their life cycles, the heavier atoms spewed into space, joining more hydrogen in the formation of the next generation of stars—or with small amounts of left-over material, planets orbiting the stars, and moons circling the planets. These solar systems, in turn, aggregated into galaxies of various configurations, depending on their ages and perhaps other circumstances of their origin. While the matter of the universe as a whole cooled and became more dispersed in accordance with the SLT, the force of gravity in compliance with the PLA drew aggregates of matter into the complex configuration of stars, planets, and moons now distributed non-randomly through space. Even as the entropy of the universe in its totality increased, so did its granular complexity.

Planets can circle their central stars for billions of years before they disperse all the energy left over from their creation; and ongoing radiogenic decay in their cores adds to this reservoir. The outward flow of this energy as heat provides the generative force for the movement of tectonic plates, whose scrapes and collisions cause earthquakes. It also heats the magma that erupts as volcanos which can build up islands or blow away mountains in a single violent event. In pursuit of an outlet, this energy reconfigures the shape of the land and the size of its oceans and seas. The process is ongoing and never ending. Mountains begin to wear down as soon as they rise up. Basins fill with their eroded surroundings, forming plains cut by rivers carrying water down its potential energy gradient from higher to lower elevations, as demanded by the SLT along

paths mandated by the PLA. The face of the land changes; the shape of the shoreline adjusts. As long as planets have energy to dissipate, they are thus in a constant state of evolution.

The creation of continents and their subsequent breakup takes millions of years. Other astronomical circumstances, like the eccentricity of a planet's orbit or the obliquity of its axis, may go through cycles lasting thousands to millions of years, generating long-term changes in climate and their associated effects on planetary topography—ice ages that cover the globe with glaciers and lower the level of the seas; then ages of global warmth, which raise the level of the oceans and submerge half the land mass of the planet. To these large-scale, long-term events, throw in the occasional astronomical catastrophe, like collision with a bolide great enough to gouge out a new moon, or a rain of meteors sufficient to speckle the planet with craters, and the vision of a globe undergoing constant change is easy to see.

Physical change is not restricted to long-term, large-scale events imperceptible to human experience, of course. Over spans of months, the seasons change from cold to warm, or from rainy to dry. Over a stretch of days to weeks, the jet streams of the upper atmosphere change their trajectories, bringing warmth and rain, cold and snow, or drought that is either hot or cold, depending on the prevailing conditions at any given spot on the planet. Ocean currents intensify or lessen, with similar effect. Where boundaries between discordant air masses become extreme, hurricanes and tornados can be generated that bring about devastating change in matters of minutes.

Regardless of the time-scale or magnitude of changes in the physical world, all of them can be traced to the fundamental forces of nature, which create gradients of mass and potential energy, then resolve those gradients in accordance with the SLT, along the specific pathways that the PLA demands. Every evolutionary event in the physical world, however, is independent of preceding events, save for the contingency of local and pre-existing conditions. Information from one event is not carried over to the next. The same cannot be said for the evolution of life, to which we now turn.

References and Notes

[1] Throughout this book, we use the term "evolutionary" in its broadest sense, meaning an ongoing trajectory of change over time in the overall nature and state of the natural world, both living and nonliving.

[2] Dinwiddie, 2008

[3] We speak of mass rather than matter deliberately, since there is not nearly enough matter to account for the mass of the universe. The difference is attributed to "dark matter" which has been quantified but not identified. In a relativistic universe, gravity acts on the mass intrinsic to, but not necessarily identical with, the atomic composition of matter.

[4] Kaufmann & Comins, 1996; Neyskens et al., 2015

[5] Smoot & Davidson, 1993

[6] Hodge, 2013

[7] Hazen, 2010

[8] *Encyclopedia Britannica* (http://www.britannica.com/EBchecked/ topic/463912/plate-tectonics)

[9] *Encyclopedia Britannica* (http://www.britannica.com/EBchecked/topic/134899/continental-drift

[10] Salthe, 2010

[11] Darwin, 1845

[12] *Encyclopedia Britannica* (http://www.britannica.com/EBchecked/topic/121560/climate)

[13] Tillery, 2002

[14] Irwin & Schulze-Makuch, 2011

Change in the Living World

4

We begin, as any discussion of the living world must, by discussing what it means to be alive. Definitions of life tend to focus on the way in which living systems differ from those that are not alive, and our definition does not depart from that tradition. In reality, of course, the living and non-living world is a continuum; atoms inside a living organism are the same as those outside it, differing only in relative proportions and the way they are bonded to one another. The forces that move mountains and generate storms—heat flow, gravity, concentration gradients, and pressure differentials—operate on and within living cells as well. In defining life, therefore, care needs to be taken to avoid the assumption that the living universe is fundamentally different from its non-living surroundings. Fundamentally, it is not.

Nonetheless, as life began to emerge in the course of time, that portion of the natural world that was alive underwent, at first, relatively subtle changes, and then, one monumental alteration that did fundamentally alter the nature of evolution itself. In explaining what happened and how it did so,

we begin with a summary statement of the end product of the transformation from a non-living to a living world.

In accordance with previous work by one of us,[1] we will define a living organism as *a material entity (1) with boundary conditions that encompass an internal environment at a lower state of entropy in thermodynamic disequilibrium with the external environment; (2) that consumes energy from the exterior to maintain the highly ordered interior and carry out intrinsic activity, and (3) is the product of near-exact replication of a predecessor by a process that is totally autonomous.*

This multi-faceted definition is necessary because so many characteristics attributable to life exist in the physical world. Clouds have structure and replicate themselves by splitting apart, but they don't have precise boundaries nor do the products of their fission necessarily resemble the cloud from which they came. An automobile uses energy from gasoline to generate intrinsic motion, but can't use the energy to repair a dented fender or replace its broken headlight. A photocopy machine can make endless copies of this page, but not by itself—somebody has to punch a few buttons. In other words, almost every property of living organisms in isolation can be matched by an analogous capability or property in the non-living world.[2] In their totality, however, living organisms are readily distinguishable from non-living entities, and their capacity for change over time has changed the very nature of evolution itself. To see how that is so, we need to begin at the beginning.

4.1 Origin of Life

The very concept of beginning implies origin at a discrete point in time. The universe in which we find ourselves, we believe, had a very precise beginning in both time and space. But life, not so much. The living world, in all probability, is a continuum with the non-living world not only in composition, but in time and place. By that is meant the fact that there is no specific place, and no precise time, at which some boundary between the non-living and the living was crossed. [3]

There is no evidence, however, to validate this assertion. It is not logically impossible that some fortuitous series of events at some specific place and time culminated suddenly in a self-contained, autonomously self-replicating cell that constituted *the singular* origin of life, from which all subsequent life devolved. Most hypotheses about the origin of life make the opposite assumption, suggesting that life emerged from its non-living surroundings by an extremely slow, erratic, and imprecise process that was widespread (possibly global), redundant, and drawn out over millions of years.[4] Logic, and the laws of nature, we believe, favor that assumption.

This book is an attempt to explain why lasting changes happen — at all times, anywhere, at all levels. It seeks to be as generic as possible. Thus, in speculating on the origin of life, our treatment will look for explanations that could apply redundantly throughout the universe and at any point in time, once all the necessary elements have been formed (in the first generation of stars, and thereafter).

The literature on the origin of life is vast, and no two scientists have the same view about how it happened or could happen. There are, however, recurrent patterns of thinking about how life could have originated wherever the components and conditions were right. Among the assumptions about which there is broad (if not quite universal) agreement are the following:

1. Complex biomolecules were synthesized prebiotically from simpler molecules. In the classical view,[5] popularized by A. I. Oparin and J. B. S. Haldane, simpler molecules like carbon dioxide and ammonia could react to form organic acids and amines when provided with an input of energy, as from naturally occurring lightning or, in the case of Stanley Miller's classic experiment,[6] an electrical discharge. Interest has always focused on carbon-based compounds, not only because life on Earth is carbon-based, but theoretical arguments make carbon by far the favored atom for constructing the backbone of larger molecules.[7]

In addition to the formation of amino acids, carboxylic acids, and other biomolecular precursors by energy pulses of ultraviolet light or electrical discharges through primitive atmospheres, as in the classic view, outer space itself is a source of biomolecular precursors. Dust clouds in interstellar space are known to be nurseries for organic compounds, delivered to Earth in the form of meteorites carrying over 40 amino acids and carboxylic acids, many of which are part of the Earth's biosphere.[8] Also suspected of contributing to primordial organic synthesis are impact shocks[9] and thermal and chemical gradients at undersea hydrothermal vents.[10]

The polymerization, or joining together of smaller biochemicals, like amino acids to form peptides, and nucleobases to form nucleic acids, has long been viewed as problematic because those reactions require dehydration (the removal of water), which ordinarily is difficult to achieve in aqueous solutions. Two ways around this problem have been suggested. The first is that dehydration-condensation reactions can occur in water if an appropriate condensing agent, like cyanimide or a polyphosphate is available,[11] and they would have been, under primitive Earth conditions. The second is that recurrent cycles of dehydration and rehydration could serve to both concentrate the reactants and promote the removal of water during their condensation.[12]

Thus, the existence of a reasonably full suite of biomolecules just prior to the dawn of life, produced by a wholly abiotic process, is now viewed as plausible.

2. Energy dissipation was both the cause and effect of increasing chemical complexity.

The construction of larger, more complex molecules from smaller, simpler ones is inherently an energy-consuming process. While the entropy of the products is reduced as they become larger and more complex, the energy consumed in their construction is degraded, so that the overall entropy of the resulting compounds and their surroundings is increased, as required by the Second Law of Thermodynamics (SLT). The synthesis of larger, organic molecules degrades more energy, not only in the process of their synthesis, but in the larger number of energy-consuming chemical reactions now made possible by these larger molecules, among themselves and with their environment, thereby providing a steeper, more efficient, and complete degradation of energy, in accordance with the Principle of Least Action (PLA).

Bear in mind that planets form by accretion due to gravity, resulting in bodies differentiated into a dense core of heavy metals that generate heat through radioactive decay, and a metallic, alkaline (reducing) crust. One scenario[13] very compatible with the theme of this book is the following: As heat from the interior drives the expulsion of volatile gases, like carbon dioxide (CO_2), hydrogen (H_2), nitrogen (N_2), nitrous oxide (NO_2), and sulfur dioxide (SO_2), an oxidation-reduction and pH gradient is set up between the reducing, alkaline crust, and the oxidizing, acidic water above into which the volatile acid-forming gases—especially CO_2—become dissolved. The resolution of these gradients results in reduction of bicarbonate (the dissolved form of CO_2) to formate, methane, and other carboxylic acids. By adding nitrogen to the carboxylic acids, amino acids form. Once the amino acids have polymerized into small peptides, they can constitute "nests" for metallic atoms which

give the polypeptides catalytic capabilities, or the tendency to promote other reactions. An incipient metabolism will have thus been created, driven by the pressure to resolve the pH gradient between the planet's alkaline, reducing crust, and the acidic, oxidizing water at its surface.

In this scenario, ocean vents are particularly attractive putative sites for protometabolic activity. Porous vents replete with minichannels could form microcompartments where protometabolic activities would have been promoted by the sequestration of interacting amino acids, nucleic acids, and lipids.[14] The close proximity of cold ocean water would have been an added asset for stabilizing compounds like methane, which could serve are energy donors for more chemical activity. The thermal gradient itself could even have driven prebiotic metabolism toward higher levels of organization.[15] Alternatively, on early planetary surfaces where water would have pooled occasionally in craters distributed in profusion across the surface, repeated cycles of hydration and desiccation, reiterated millions of times globally, could have provided a vast number of protometabolic "experiments" over time. From some of the experiments, certain recurrent forms would likely have emerged with greater frequency and success.[16] Eventually, cells encapsulated within lipid membranes would have been able to float free of their microchamber nurseries or their condensation-rehydration pools and become protocells.

Reactions inside these protocells would have proceeded at a rate much higher than they did in the unsequestered surrounding environment, partly because the higher concentration of reactants inside the protocell favored more frequent reactions, but also because the growing repertoire of biomolecules with catalytic activity, however primitive, promoted reactions at a much higher rate inside the vesicles than outside. Protocells thus would have provided a much higher rate and more efficient utilization of energy than the surrounding environment, increasing entropy faster and to a greater extent as mandated by the PLA, in the direction required by the SLT.

The emergence of protometabolism on any planetary body represents a watershed event for energy processing on that world. The miniaturization of energy transformations brings the preponderance of energy flow and entropy production down to the molecular level. At that point, more energy per unit mass is being degraded by the interaction of molecules than the erosion of mountains. Physics has given way to chemistry as the prevalent venue for energy degradation.

Yet a barrier persists in keeping us from designating this protometabolic ensemble as being alive. However intricate or efficient the energy flow through molecular interactions becomes, the reactions remain chance events from one episode to the next, unless the probability of their recurrence can be enhanced. The little packets of chemistry that degrade energy with such vigor have to have a way of replicating themselves, to ensure that their effectiveness is propagated through time.

3. Replication of form and function with high fidelity became autonomous. Before a repertoire of reproducible metabolic interactions based on replicable molecules could have been considered to be alive by our definition, that repertoire would have to have become encapsulated into structures capable of replicating themselves autonomously. Once carboxylic acids or other hydrophobic (water-repelling) molecules became abundant enough, self-forming vesicles enclosing protometabolic activity would have been likely, just as individual drops of detergent form readily in dishwater. But the ability to replicate themselves autonomously, with the descendent packets of metabolism being virtually identical to their predecessor, is more difficult to imagine.

Structure-creating and energy-transforming metabolic reactions, along with the means for replication or reconstitution of the molecular machinery for those processes, were very likely a matter of stochastic probability in the eons leading up to the onset of what Chaisson[17] calls the Life Era. Every new protocellular vesicle spontaneously forming or budding off from a predecessor, or each new microincubation chamber that bubbled forth in porous rock, may have contained a collection of catalytic and replication-prone molecules, but the collection would presumably have been composed of only a statistical approximation of the composition of other protocells or microchambers around them,

both before and after their ephemeral existence. Some may have been endowed with a particularly favorable complement of molecular components, and those particular protocells may have survived longer than others. But they too would have passed out of existence with no lasting effect, *unless and until* a mechanism had emerged for ensuring the reproducibility of that particular favorable set of components.

One can envision two possible ways in which the contents of a descendent protocell could have been generated as an exact copy of its predecessor. One would have been a selectivity mechanism which admitted into the protocell's interior only those components present in the predecessor. The other would have been a biosynthetic capability which recreated the same components as in the predecessor. While the first mechanism may have operated to some degree, especially for admitting or excluding smaller metabolites, the larger molecules with catalytic capabilities would have been much more likely to reliably appear as a consequence of being manufactured on demand. Both mechanisms require the ability to store and transmit a new commodity: information. The life era truly began only when the means for carrying out processes in a reliable and reproducible way became encoded in a replicable form of information. *Thus was introduced into the evolutionary process a primary role for information,* superimposed as a value-added constraint on the laws of chemistry and physics within which all change occurs.

Life on Earth today encodes the information necessary for all living processes, including the ability to self-replicate, in hugely long double-stranded molecules of deoxyribonucleic acid, or DNA. Almost no biologist, however, believes that DNA was the first informational molecule capable of being replicated at the early stages of the Life Era. Ribonucleic acid, or RNA, has all the informational capacity that DNA has, but operates through shorter, single-stranded segments, and furthermore has catalytic properties. A widely held view, therefore, is that an "RNA World" must have preceded the age of DNA-based cellular control.[18]

Two questions remain controversial about the RNA World. The first derives from the fact that

RNA itself is a complicated molecule and seems highly unlikely to have emerged in its current form from the primitive metabolic environment that must have characterized the earliest, incipient forms of replication. A more primitive mode of replication is almost sure to have preceded the process that took hold in the RNA World.[19] The second question is whether any complex, replicating molecule could have arisen before a fairly complex repertoire of metabolic reactions had evolved. This question, in oversimplified form, is: What came first – replication or metabolism?

The trivial answer is that any replication requires a metabolic process of some sort, but the persistence of any *particular* metabolic pathway—and reliable replication depends on the persistence of particular metabolic pathways—requires some means of replication. The circular reasoning is obvious. As in the case of the chicken and the egg, it is difficult to see how one could emerge without the other. Therefore, while arguments for replication first[20] on the one hand, and metabolism first[21] on the other, have been made, those models[22] that envision coevolution of metabolism and replication are most likely in our view to represent the actual history of the origin of life. The centrality of the evolution of metabolism to the appearance of life gives credence to our thesis that energy dissipation under the SLT and PLA was the guiding and driving force behind the origin of life.

Persuasive arguments have been made that RNA is the evolutionary remnant of an earlier inorganic template, serving a cruder alignment and catalytic function found in the pre-life era. Minerals composed of reproducible crystalline architecture could have served such a non-living template function, and the metallic atoms that are co-factors in enzymes today may be descendant reflections of catalysis performed by the naked atoms before the dawn of life.

Graham Cairns-Smith provided the most complete scenario for evolution of replication and catalysis originating from crystalline minerals.[23] In his view, the crystalline structure of fine-grained clay could have provided the earliest templates and catalysts for organic synthesis, by providing surfaces that adsorbed amino acid and nucleoside monomers on their surface in a particular orientation that made polymerization more likely. Energy flow from the environment, like ultraviolet radiation or heat, or chemical energy donors like polyphosphates (possibly generated through a primitive photosynthetic capacity in certain crystalline structures), could have provided the energy necessary to drive the construction of oligopeptides and oligonucleotides. Interaction between these two molecular species could have mutually supported their co-evolution.[24] Experiments have shown that RNA is stabilized by adsorption to clay surfaces, and still is able to serve as a template for replication.[25]

Assuming that replication was guided initially by mineral templates, this process cannot be said to be autonomous until a mechanism emerged that could operate without the external mineral template. In light of these difficulties, the most plausible models for the origin of autonomous replication envision that metabolism, replication, differentiation and division all began together in protocellular structures.[26]

It should be emphasized at this point that the pre-life era—when the chemical components of what would but had not yet become living systems were being assembled—represented a significant evolutionary change in the way energy was being processed in nature. Well before living cells appeared, nature was using the building blocks of later life to degrade potential energy in planetary pH and oxidation-reduction gradients.[27] That life would eventually emerge from these chemical innovations is interesting to us as living organisms; but it could and should be argued that they emerged with no inherent drive toward creating life, but rather were nature's way of resolving physicochemical asymmetries created by the way in which planets were formed.

4.2 Evolution of Life

The evolution of life is an extension of the trend of ongoing change in the natural world that began at the moment the universe came into existence. But the nature of change in the living world differs from change in the physical world in several important ways.

First, the scale and precision at which meaningful change occurs is much, much smaller in the living than in the non-living world. While a shifting grain of soil or a blowing leaf in the wind will change the face of a mountain in a miniscule, objective sense, they have no significant impact on the shape of the mountain in themselves. A point mutation in a chromosome, on the other hand, can change the readout of a gene product, and thereby affect the shape or function of an entire organism. The course of a mountain stream can shift its path by shortcuts measured in meters to kilometers, to carry water to a lower level of potential energy by a more direct path than previously. A metabolic pathway, through introduction of a slightly altered catalytic capability in just one critical enzyme, can likewise introduce a more efficient way to extract energy from a metabolite; but the relevant change lies well below the resolving power of the human eye.

Secondly, change occurs much faster in organic evolution than it does in the geophysical rearrangement of the planet. Continental plate movements take millions of years to be discernible. Mountains take hundreds of thousands of years to rise up and erode away. While organisms and ecosystems can remain static under persistently stable environmental conditions for long periods of time, punctuated changes in the evolution of species brought about by environmental changes can occur within thousands to just hundreds of years. Human activities, like industrial pollution and the use of insecticides, can induce noticeable changes in morphology and survivability of some species within decades. Experimentally, significant changes can be induced in a matter of weeks in fruit-flies and days in microorganisms.[28] By way of contrast, consider the breakdown of rock and soil into its elementary chemical constituents. This process occurs naturally through weathering or other abiotic forces, but happens by orders of magnitude faster through the action of bacteria and fungi.[29] While non-living phenomena like snow storms and tornados do occur over time scales of minutes to days, they are not lasting changes in the weather, and therefore lie outside the meaning of evolution as we have defined it.

Thirdly, and most innovatively, biological change occurs through and among a constellation of individual entities (organisms) that survive for but an instant on geological time scales but are replaced recurrently by near exact copies of themselves, perpetuating the collective—be it a gene, species, higher taxon, or entire ecosystem. Every new generation of an organism faces the challenge of surviving long enough to reproduce its kind, and pass along the genetic information on which its survivability is based. Once an informational system had arisen that could direct the exact replication of structural and catalytic molecules, and a mechanism had emerged for replicating and passing on to successor cells those informational components, the stage was set for the onset of a new force in evolution: the principle of natural selection.

4.3 Emergence of Self-Perpetuation and Natural Selection

In its classic form as advanced by Charles Darwin,[30] the unit of natural selection was the individual organism. Those organisms that chance modifications endowed with a greater likelihood of survival in the struggle for existence were more likely to reproduce than other members of their species which were less fortunately endowed. Darwin assumed the existence of information in some form that passed through successive generations, but did not know what it was or that it would come to be called a "gene." What he did know was that the transmission of information from one generation to the next enabled favorable modifications to persist in proportionately greater numbers over time.

In the neo-Darwinian formulations of the early 20th century, emphasis shifted to the population gene pool as the level at which natural selection could best be described.[31] Those alleles, or variants of a particular gene, that conferred an adaptive advantage were more likely to increase proportionately in the entire gene pool over time. With elucidation of the molecular basis of genetic information in the second half of the 20th century, views about the unit of natural selection became blurred, with emphasis ranging from minimal nucleotide sequences composing small segments of specific genes, to whole taxonomic units at the species, or even higher, level. In the case of the former, went the argument, what actually conveys adaptive benefit is the portion of the gene that critically affects the function of a gene product (or regulatory process) in a more or less favorable way. In the case of the latter, support comes from broad tendencies in the evolution of whole groups, such as the differential survival of some species over others.[32] Regardless of the focus of the unit of natural selection, what they all share is the fact that some biological features are more likely to survive long enough for the informational program that directs their construction and function to be passed along in greater numbers to succeeding generations. Perpetuation through reproductive success, for whatever reason, is the ultimate arbiter of which changes persist through time in the living world.

4.4 What Is Selected by Natural Selection?

A biological adaptation is a feature that increases the probability of survival. Organisms achieve reproductive success through adaptation to their circumstances and environment. Since reproductive success is the end point of natural selection, adaptation is what is selected for by natural selection. But what makes an adaptation "adaptive"? How, or in what way, do superior adaptations better fit their organisms for survival?

For Darwin, natural selection tended toward greater organization, ". . . inasmuch as the parts are thus enabled to perform their functions more efficiently." Darwin used the words "efficient" or "effective" twenty times to describe the effects of adaptation.[33] Over a century ago, Lawrence Henderson wrote that ". . . beneath adaptations, peculiar and unsuspected relationships exist between the properties of matter and the phenomena of life."[34] Among those properties of matter, special attention has focused on energy flow and its utilization, since energy is obviously integral to the survival of living systems. Alfred Lotka suggested the importance of energy processing, when he wrote early in the 20th century that "natural selection will so operate as to increase the total mass of the organic system, to increase the rate of circulation of matter through the system, and *to increase the total energy flux through the system* (italics added). . ."[35].

The topic of energy flow inevitably leads to thermodynamics, and from there to entropy. The modern trend toward viewing life through a thermodynamic lens is generally regarded as dating from Erwin Schrödinger's pronouncement that organisms maintain the living state by drawing negative entropy (negentropy) from their environments.[36] For decades, Harold Morowitz provided compelling arguments for the fact that biological organization is driven by the flow of energy, with living systems maintaining their low level of entropy at the expense of a net increase in entropy in the system and the environment of which they are a part.[37] Eric Chaisson has been explicit in linking the SLT to organic evolution, arguing that the ongoing expansion of the universe in accordance with the SLT is the ultimate engine for the evolution of life.[38] He has advanced the concept of "energy rate density" as a comparative metric for the degree to which energy is degraded by different systems.[39] Energy rate density is the amount of energy transformed per unit of time and mass. A comparison of energy rate densities makes clear that living systems degrade energy much more effectively than purely physical systems.[40]

Recently, Arto Annila and his colleagues have taken the expression of evolution in thermodynamic terms to a new level of formalism and sophistication.[41] They have modeled evolution as an "energy density landscape in flattening motion." When written as a differential, evolutionary motion is described as a pathway for the steepest descent in energy. When written as an integral, evolutionary motion is pictured to take place over the shortest path. These are clear expressions of the PLA, thus bringing natural selection and the PLA into synchrony. By applying their models to complete ecosystems,[42] these authors further provide an explanation for macroevolutionary processes as well in terms of the PLA.

Sharma and Annila have argued specifically with regard to natural selection that "the rate of entropy production by various mechanisms is the fitness criterion of natural selection."[43] We disagree, in the sense that the fitness criterion of natural selection has to be reproductive success. But we take Annila's deeper meaning to be that natural selection leads to reproductive success because it favors those organisms that can best maintain their negentropy by maximizing the entropy of the overall system of which they are a part. Since a higher rate of entropy production comes about by degrading more energy, more quickly, per unit of mass, Annila goes beyond previous authors in relating fitness to the degree to which organisms *degrade energy most effectively*, in accordance with the PLA. An organism that can derive more energy from a unit mass of resources more quickly has an advantage over its competitors, and therefore is favored by natural selection.[44] From this perspective, **the answer to the question of *why* evolution occurs in the living world is that natural selection favors the emergence of adaptations that make ever more efficient uses of energy, thereby maximizing its degradation and increasing entropy overall**.

While this may explain in thermodynamic terms the engine that powers living processes, an explanation of how and why these manifestations of the SLT and the PLA in living systems lead to a proliferation over time of diverse forms, some of which become increasingly complex, requires further elaboration. As a backdrop for this discussion, we will next consider the major episodes in the history of life on our planet.

4.5 Major Transitions in the History of Life on Earth

The course of evolutionary change on Earth has been more like a series of cascades than a smooth flowing stream. Between the abrupt and substantial changes (innovations) represented by waterfalls, have been long periods of stasis—the course of evolution undergoing relatively minor changes in the level and direction of flow of the stream of life. The concept of punctuated equilibrium, originated and popularized by Niles Eldredge and Stephen J. Gould,[45] is the scientific version of this metaphor, and it dominates contemporary thought about the tempo of evolution.

Punctuated equilibrium operates at a number of levels, varies in magnitude, and has fragmented life into millions if not billions of different directions. But a few overarching transitions have lifted the course of evolution to a broad new plateau, characterized by increased energy flow densities by orders of magnitude at each new level (Fig. 4.1). Among those major transitions, most biologists would agree on at least the following list:

Unicellular Chemotrophs. Stable, reliably functioning, consistently replicable cells eventually emerged from what were likely a huge number of protocellular fits and starts (Fig. 4.1a). They subsisted on chemical energy from their environment, with the metabolic machinery for extracting that energy from inorganic compounds (Fig. 4.1b). The trick was to extract hydrogen atoms (H) from an inorganic source, such as hydrogen sulfide (H_2S), and use them in two ways. The first was to transfer them to molecules with lower chemical potential energy, while capturing the energy difference in the form of high-energy compounds like adenosine triphosphate (ATP) and nicotinamide adenine dinucleotide (NADH). Not only did this retain the energy in a more useful form than just heat, it retained the energy until it was needed, and thus could be used more effectively by the cell. The second use of H was to reduce carbon dioxide (CO_2), as a means of producing the organic compounds ($-CH_n$) on which living processes rely. The metabolic machinery for carrying out these reactions must have emerged at the earliest stages of life, possibly from elaborately specified mineral configurations that were taken up and improved upon by organic macromolecular catalysts.[46]

Soon enough—perhaps almost from the beginning—some cells developed the ability to extract hydrogen from organic compounds like methane (CH_4) as well. Of course, other organisms that synthesized organic compounds could serve as food also. Those that manufacture their own food from inorganic substrates are autotrophs ("self-feeding"), while those that subsist on other organisms are heterotrophs ("other-feeding"). Either way, the end point of these energy yielding pathways was primarily organic acids, like pyruvate and acetate, and alcohols, like ethanol, which were excreted into the environment. The pathway ending in pyruvate is known as glycolysis. The pathway producing alcohol is called fermentation.

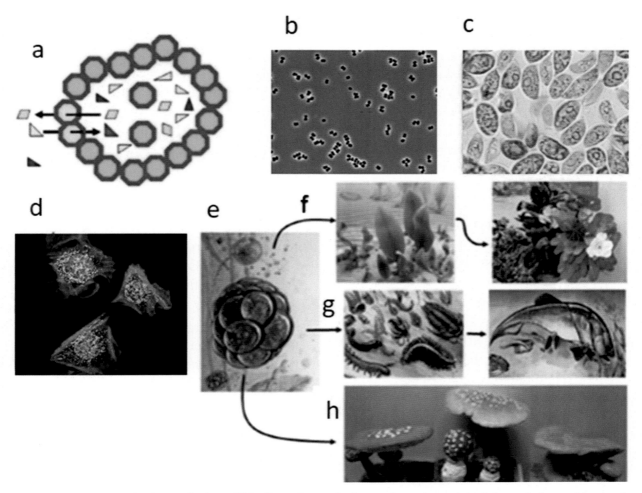

Fig. 4.1. Major steps in the evolution of life. Several watershed stages have characterized the evolutionary history of life on Earth, including: (a) the appearance of protocells with minimal metabolism; (b) simple prokaryotic cells deriving energy from their chemical surroundings; (c) unicellular microbes with green pigments of chlorophyll to capture solar energy; (d) symbiotic eukaryotic cells with specialized organelles derived from ingested prokaryotes, and a membrane enclosed nucleus. The early eukaryotes diverged (e) into multicellular forms leading to (f) autotrophic plants, (g) heterotrophic animals, and (h) heterotrophic fungi. (Arrows are not meant to imply direct ancestor-descendant relationships). *(a) art by Louis Irwin; (b) NASA/JPL-Caltech; (c) Elena Lopez Peredo, NSF; (d) Dylan Burnette and Jennifer Lippincott-Schwartz, Eunice Kennedy Shriver National Institute of Child Health and Human Development, NIH; (e-h) photography by Louis Irwin of exhibits at Denver Museum of Nature and Science, with permission.*

They were simple cells, with no internal membranes, complex architecture, or much functional specialization. Lacking an organized nucleus, we call them *prokaryotes* ("before nuclei"). They multiplied by splitting apart after replicating their genetic information stored in a single chromosome, so that each daughter cell received copies of the instructions for synthesizing the proteins that directed their assembly, catalyzed their metabolic pathways, and orchestrated the replication and cell division process. And they were microscopic in size. But they markedly increased the transformation of energy compared to their surroundings. While physical planetary processes were using energy at an average rate of 75 ergs per second per gram (ergs/s/g),[47] the simplest prokaryotic cell at a hydrothermal vent has been estimated to process around 200 ergs/s/g.

Unicellular Phototrophs. Early in the history of life, certain unicellular autotrophs developed the ability to harvest the virtually inexhaustible energy of sunlight for producing their own food (Fig. 4.1c). Specialized protein-metal complexes like chlorophyll evolved as molecules with loosely-held electrons that could be bumped to higher energy levels by absorbing photons of light. As they fell quickly back to their ground state, the electrons gave up energy that could be captured to generate ATP and NADPH.[48] In the earliest evolutionary stages, cells like modern-day green sulfur bacteria extracted H atoms from H_2S and from other organic compounds, for the reductive synthetic pathways pioneered by the chemotrophs. The most elementary of photosynthetic mechanisms may even have evolved from clays that provided photochemical machinery for making a few metastable molecules, such as polyphosphates, very simple carbohydrates, and ammonia.[49] Eventually, the geochemically produced inorganic and organic sources of H became depleted. A more sophisticated photosynthetic mechanism then evolved which could couple electron excitation with the removal of H from H_2O, a nearly ubiquitous resource.[50]

Absorption of photons and the splitting of water degraded the energy from the sun and increased entropy. Living cells became a new way to dissipate the ubiquitous solar energy impinging upon the Earth. Now the energy of sunlight could be degraded, not only through absorption by the substrate which dissipated it as heat, or absorption by water which dissipated it through evaporation, but by mechanisms within living cells that used it to build up organic molecules at the expense of a large increase in entropy and a substantially more effective use of solar energy.

Among the earliest cells to master phototrophy were the cyanobacteria, which made use of a blue-green pigment optimally arranged for capturing energy from the particular wavelengths which dominate sunlight. These prokaryotes were probably the primary producers of free oxygen in the early history of the Earth, and they continue to pour free oxygen into the Earth's biosphere, over 3.5 billion years after their first appearance on the scene.

Light in itself does not yield that much more energy than inorganic chemicals. One photon of light delivers about 2 electron volts (eV), compared to the oxidation of one atom of hydrogen by iron, which yields about 1.6 eV.[51] The real impact on energy utilization came from the metabolism made possible by the photosynthetic release of free oxygen (O_2).

Fig. 4.2. Oxidative metabolism. The evolution of photosynthesis made possible the use of the ubiquitous energy source of sunlight, and globally available carbon dioxide (CO_2) and water (H_2O) for the synthesis of high-energy glucose ($C_6H_{12}O_6$) in plants. The oxygen (O_2) thus generated could then be used by both plants and animals to oxidize the nutrients in energy-yielding steps, captured as ATP, back down to H_2O and CO_2. It became particularly important after life colonized the land, where O_2 was much more concentrated in air than in water, but the metabolic steps originated much earlier. *Graphic design by Louis Irwin; photo by J.S. Peterson, USDA NRCS PLANTS Database (left), and the National Zoo, Smithsonian (right).*

Oxidative metabolism. Oxygen is very reactive, and the first thing it did upon being released from photosynthesizing cells was to oxidize whatever substrates were available. These were mostly reduced metals, so the earliest global evidence of biogenic oxygen production was the appearance of red bands of oxidized iron minerals around the world. Once all the minerals that could be oxidized had reacted with oxygen, the balance of oxygen being produced through photosynthesis began to accumulate in the atmosphere. This was bad news for most living organisms, for which oxygen would have been a potent toxin. But the filter of natural selection resulted in a few varieties of cells that emerged in time because of their ability to detoxify the oxygen by reducing it back to water. A side benefit was the fact that water has a much lower chemical potential energy than does pyruvate, lactate, or ethanol, the end points of metabolism possible in the absence of oxygen. That meant that oxidation of H all the way down to water released much more total energy than was possible with organic compounds as the final H acceptors. This is the essence and revolutionary significance of oxidative metabolism.

Organic materials, like wood, have a tremendous amount of energy locked into their C-C and C-H bonds. That energy is released as light and heat when wood is burned in a fireplace or campfire. The reason so much energy is released from burning wood is that all those H atoms in cellulose and other organic constituents are stripped by oxygen in the process of combustion and transferred directly to water (with carbon dioxide being the other product).

The way that oxygen came to be rendered less destructive to living cells was by the innovation of transferring H down its potential energy gradient in small steps at a time, so that no one step released enough heat or ruptured enough vital chemical bonds to be damaging.[52] Instead of a single energy cascade the size of Niagara Falls, hydrogen dropped down through a series of smaller cascades that lowered its potential for damage to a minimum at each step, till a final mild reaction added H to O to reach the metabolic endpoint of H_2O. As was the case with the earlier chemotrophs, some of the energy released by the movement of hydrogen to lower levels of potential energy was used to form high-energy phosphate bonds in molecules like ATP, where the energy could be stored until needed (Fig. 4.2).

Oxidative metabolism is not essential for life. Except for a few animal groups with very high metabolic rates, like mammals and birds, most organisms can survive for long periods of time without calling on the high energy yield of oxidative metabolism. Nutrients have to be reduced, to be sure, to obtain energy, but the metabolic cascade can stop well short of the final oxidation of hydrogen to water. The metabolic process of glycolysis, for instance, converts one molecule of glucose to two molecules of pyruvate, releasing enough energy to provide a net gain of two energy-storing molecules of ATP in the process. Another pathway, fermentation, converts a molecule of glucose into two molecules of ethanol, releasing two molecules of CO_2 and liberating about the same amount of energy as glycolysis. Fermentation and glycolysis can proceed in the absence of oxygen. But if oxygen is present, most organisms can further degrade the end products of glycolysis and fermentation all the way down to six molecules of CO_2 and up to 36 molecules of ATP, or 18 times the amount of energy that can be obtained by glycolysis or fermentation alone.[53] The capacity that oxidative metabolism has to extract the maximum amount of energy possible clearly provides an advantage to those organisms that possess the metabolic machinery for doing so. And maximal degradation of energy is under constant pressure from the PLA.

Unicellular eukaryotes. For 1.5 to 2.0 billion years – about half of the entire history of life on Earth – life persisted in the form of simple, prokaryotic cells. A membrane constructed of two layers of hydrophobic (water-repelling) lipid molecules separated the aqueous interior of the cell from its aqueous external environment. The membrane was interspersed with protein channels for admitting nutrients from the environment and expelling waste products from the interior, as well as serving as a conduit for the protons that diffused down their concentration gradients into the cell to drive the formation of ATP by chemiosmosis. Proteins attached to the inner side of the membrane grabbed the circular strand of

DNA that carried the cell's genetic information and transcribed it into RNA, which was then translated by other protein complexes into different proteins. A sturdy cell wall surrounded the microscopic organism, protecting its fragile contents from osmotic swelling.

Many variations on the theme of these simple cells came and went, as natural selection optimized the mechanisms for energy degradation and self-replication. But the basic structure and strategy of the living cell remained unchanged for close to two billion (2 x 1000 x 1000 x 1000) years. One major division did come about early on. Since those cells which could produce their own food by harvesting chemical or light energy could serve as food for those that could only derive energy from other organic compounds, the living world was divided between autotrophs and heterotrophs, the latter providing yet another, more complete step of energy degradation. This process, whereby some cells lived by consuming other cells, set the stage for one of evolution's most significant innovations: the eukaryotic cell.

Natural selection almost surely engaged in a vast number of trial-and-error experiments in the process of revolutionizing cellular life. What we see in the eukaryotic cell of today is the culmination of a number of innovations, each contributing to what arguably was the single most significant evolutionary change in the history of life. Those innovations include the following:

1. Loss of cell wall, allowing the cell to become larger and more complex.

2. Formation of internal cytoskeleton, to provide structural stability, since a cell wall was no longer present to prevent osmotic swelling (influx of water, due to higher internal concentration of dissolved contents). The internal cytoskeleton was made of tubulin and actin — two contractile proteins that would subsequently by used for moving parts of the cell, like the spindle fibers that move chromosomes, and eventually whole cells. The contractile properties of muscle are based on the early evolution of contractile proteins.

3. Infolding of external membranes, to increase surface area for nutrient absorption.

4. Pinching off of infolded membranes, forming enclosed internal compartments, or organelles. One class of such membranes retained attachment of protein-RNA complexes, or ribosomes, for synthesizing proteins. Internal membranes studded with ribosomes became the endoplasmic reticulum.

5. Formation of digestive vesicles, evolving into lysosomes.

6. Endocytosis of prokaryotes with specialized metabolic capabilities by another prokaryote lacking or less proficient in those metabolic processes. Endocytosis is the process by which a cell engulfs another cell from the exterior. Loss of the cell wall made endocytosis possible. The resulting symbiotic (mutually beneficial) union was good for both the host and the endosymbiont. *Mitochondria* came from prokaryotes specialized for generating ATP through oxidative metabolism. *Chloroplasts* originated from cyanobacteria which had mastered the process of photosynthesis. *Peroxisomes* came from prokaryotes with detoxifying capabilities. The endosymbiotic origin of cell organelles as envisioned and advanced by Lynn Margulis[55] is one of the great insights into the history of biological evolution.

7. Formation of cilia and flagella, enabling the onset of cellular motility and ability to generate motion in an organism's surrounding.

8. Enclosure of DNA within a membrane-bounded vesicle, resulting in a primordial nucleus (eukaryote = "true nucleus"). Within the nucleus, genetic material was transformed from a circular structure to multiple rod-shaped extensions, or chromosomes, which allowed for genomic expansion and multiple starting sites for replication and transcription.

9. Incorporation of a spindle apparatus for separating chromosomes. In time, mechanisms for dividing up genetic information, now localized in linear chromosomes, evolved — quite possibly through endosymbiosis of other prokaryotes that donated contractile proteins to the spindle apparatus for separating chromosomes during mitosis and meiosis.[56] The origin of haploid, or single, copies of the genome then made possible the fusion of two haploid cells to produce diploid cells containing double copies of genetic material— the essence of sexual reproduction. This vastly expanded the organism's ability to innovate, due to the multiplication of different gene combinations resulting from the sexual union of two different copies of the genome, each with variant forms (alleles) of different genes.

The complicated nature of the cell division process, along with the other specializations provided by each particular combination of prokaryotic component parts, led to stabilization of form and function in succeeding progeny. Horizontal gene transfer, as commonly practiced among prokaryotes, became much less prevalent, and the concept of the species, as an elongated succession of like, interbreeding organisms, became a firmly entrenched characteristic of all eukaryotes.

The eukaryotic cell is up to a thousand times larger than a prokaryotic cell, and is vastly more complex. The number of protein-coding genes is 40% greater in the simplest unicellular eukaryote (yeast), compared to a bacterium like *Escherichia coli*. The number of genes in simple colonial eukaryotes, like green algae and social ameba is more than three times greater. The larger size and complexity are also manifested in terms of more effective energy degradation. As noted earlier, a prokaryotic cell at a hydrothermal vent degrades an estimated 200 ergs/s/g, but simple protists (unicellular eukaryotes) utilize close to 1000 ergs/s/g, or five times as much.[39]

Multicellularity. The assembly of eukaryotes from multiple combinations of specialized prokaryotes gave rise to many different types of eukaryotic cells. Some, with a rich endowment of chloroplasts, could produce their own food by photosynthesis, and became something like modern green algae (Fig. 4.1f). Others good at oxidative metabolism but not at photosynthesis could efficiently extract energy from other organic compounds, but had to obtain those starting compounds from their environment. Among the latter, some simply ingested dissolved organic material, like modern fungi such as yeasts, while others consumed intact organisms as food, much as today's ameba and paramecia ingest other living organisms or parts thereof.

These variable collections of unicellular organisms are collectively known as protists. Most biologists today recognize them, somewhat

reluctantly, as a distinct Kingdom of life. The reluctance has to do with the fact that they most likely had multiple origins (were polyphyletic). Also, a clear distinction between unicellular species and multicellular successors possessing many of the characteristics of their predecessors, is often lacking. Colonial organisms, like green algae, for instance, presumably gave rise to plants (Fig. 4.1f). Heterotrophic protists that became good at ingesting other organisms, then digesting the food internally, evolved into animals (Fig. 4.1g). Other heterotrophs that absorbed dissolved nutrients originating from other organisms, became fungi in their multicellular form (Fig. 4.1h).

The tendency to form larger, multicellular aggregates, evolved numerous times in different groups of organisms. While the most common classification scheme today recognizes three kingdoms of multicellular eukaryotes — plants, fungi, and animals — each of those kingdoms is probably polyphyletic; that is, different groups within each kingdom probably had independent evolutionary origins among the protists.

Whatever their family lineages, it seems reasonable to assume that multicellularity itself originated simply enough when dividing unicellular organisms failed to separate. Natural selection under certain circumstances favored the survival of such colonial aggregates. In a subsequent stage, subgroups of cells which tended to specialize in certain functions, provided the group with more effective and efficient means of carrying out processes vital for survival. Cellular differentiation became a feature of the organism's life history, and a developmental time course became a definitive characteristic of every multicellular organism. This, of course meant that programming for developmental transitions had to become part of the genetic endowment of each species. The need to provide information for the construction and function of each stage of the life cycle in multicellular organisms is reflected in the larger genomes of multicellular, as opposed to unicellular, organisms. Protein-coding genes number 6,696 for the unicellular yeast, while the number of genes that are translated into proteins in multicellular organisms is 9,820 for bread mold,

14,076 for the fruit fly, 23,062 for the mouse, and 57,995 for rice. [39]

Metabolic rates, as measured by O_2 consumption per mass of organism per unit of time, are roughly an order of magnitude greater in multicellular than in unicellular organisms.[57] It appears that a critical amount of O_2 had to accumulate in the atmosphere before multicellularity could gain a competitive foothold. This apparently occurred between 800 and 600 million years ago, as this was when the first evidence of macroscopic, multicellular organisms began to appear. Simple, sessile marine organisms were the first to achieve notable size. These heterotrophs were the forerunners of the animal kingdom.

Air holds much more oxygen than water, so full advantage of oxidative metabolism was not achieved until some organisms adapted to life on the land. With abundant oxygen and sunlight in this environment, algae-like organisms could evolve into plants. Able to manufacture their own food from CO_2, and with access to the boundless energy provided by sunlight, these organisms were able to grow to large size, producing the dense moss, fern, and coniferous forests of the Devonian Period, over 400 million years ago. Once autotrophic producers of organic nutrients became established on land, opportunities for land invasion by the heterotrophic fungi and animals opened up as well. Terrestrial ecosystems complete with plants, fungi, and animals were well established by mid-Devonian times, about 475 million years ago.

Animals are multicellular heterotrophs. They begin appearing in the fossil record between 600 and 800 million years ago, in three forms. First, the sponges were multicellular, ciliated organisms that arranged themselves into stationary colonies of cells with internally-directed cilia that beat in a direction that drew water and its contents from external seawater into the animal's interior, where food particles could be collected and digested. Secondly, cnidarians like jellyfish and sea anemones developed in the form of simple radially-symmetrical bodies with filamentous or arm-like extensions which could capture passing bits of living or dead nutrients, then draw them into the organism's interior for digestion. Thirdly, worms emerged with the ability to move themselves by hydrostatic extensions and contractions of their bodies through the food-rich substrate in which they lived, ingesting their food as they moved, and digesting it internally. By 600 million years ago, worms had begun to evolve into a variety of body forms, which included appendages for grasping and locomotion, as well as antennae and other sensory organs.

Multicellularity stepped up the rate of energy consumption by living organisms. Seed plants of the type that became established during the Devonian utilize several thousand ergs/g/sec, compared to a thousand or fewer ergs/g/s by a complex protist. With the evolution of mobile animals, energy rate densities increased to the range of 10^4 to 10^5 ergs/g/s. Increments approaching an order of magnitude can thus be seen at each of these new evolutionary plateaus.[38]

While not part of their formal definition, animals are all mobile. Sponges and many cnidarians are stationary but have the ability to move their body parts and/or the medium in which they live. Worms and all their descendant forms have true mobility. The ability to move toward and capture food, or move away from and escape capture, signaled the onset of a co-evolutionary contest between predators and prey that continues to this day among animals. Even those animals that feed only on plants move about to gather or graze on their food. The evolution of whole-body mobility was a major step in the evolution of life, with significant energetic consequences.

4.6 Evolution of Diversity and Complexity in the Living World

To reiterate a central question with which this work is concerned: Why has the universe evolved local pockets of increased complexity in the face of an expanding and increasingly disordered universe overall? In this section, we are specifically concerned with the question of why organic evolution has resulted in a greater diversity of increasingly complex living organisms.

Diversity and complexity are not the same, but they are related. We will first discuss why diversity arises, then explain how diversity necessarily leads to complexity.

At an intuitive level, the origins of biodiversity and biocomplexity would appear to be straightforward: If life had a singular, simple origin, the factors that cause change and create diversity over time necessarily resulted in a greater variety of forms with increasing complexity because those were the only directions possible. Several authors have gone beyond intuitive thinking in their attempts to provide a more formal explanation for the apparent inevitability of increasing biodiversity over time. Chaisson[16] sees the expansion of the universe as the agent for generating local increases in entropy that provide the energy gradients that drive change. Sharma and Annila[58] imply that the increasing entropy associated with the "spontaneous drive for differentiation and motion towards molecular diversity" provides the motive force for generating organismic diversity. Brooks and Wiley[59] attribute increasing diversity to the inherent tendency of genetic and organizational information of living organisms to undergo alteration through mutation and genetic recombination — an increase in entropy consistent with the SLT. The tendency for diversity to increase was seen by McShea and Brandon to be pervasive enough to be formalized as the "Zero-Force Evolutionary Law"—with rising diversity and complexity, rather than the steady state, being the null expectation—all within the mandate of the SLT. While Sharma and Annila see entropy production as the fitness criterion for natural selection, the other authors argue either explicitly or by inference that evolutionary change is inevitable, entirely constrained by natural selection.

Though the assertion that diversity has increased in the living world since life began would appear to be self-evident, the facts paint a slightly more nuanced picture. Diversification has not always and inexorably increased, as the formal theories would seem to predict. Since there is little fossil evidence pertaining to the first 2.5 billion years of life on Earth, we cannot state with assurance much about the diversity of the microbial life that inhabited those ancient seas. Once fossils did begin to accumulate about a billion years ago, however, we can document the course of biodiversity in a number of categories.

The time course for growth in the number of Families[61] of marine organisms is shown in Fig. 4.3. Following a steady increase for over a hundred million years, the number of different families of marine organisms levelled off, then started to decline about 330 Mya until reaching a low point at the Permian-Triassic (P-T) boundary, known to be the greatest extinction event in the history of life. Marine biodiversity started to rise again after that, but was slowed with a second extinction event, at the Cretaceous-Tertiary (K-T) boundary, before resuming another but decelerating increase.

Terrestrial plants evolved from marine aquatic photosynthetic ancestors, and obtained a foothold on the land during the Silurian Period. By the start of the Devonian, the first ferns (Pteridophytes) were diversifying across the land in a steady increase until the late Permian, when the aforementioned crisis caused their number of species, along with those of their descendant conifers (Gymnosperms), to crash (Fig. 4.4). After the P-T boundary, conifers began to recover, but the ferns never again regained their peak diversity; and once the flowering plants (Angiosperms) evolved, they grew in diversity as the variety of conifers declined. The number of families of non-flowering plants actually peaked in the early Permian, never to be equaled again.

The first animals on land were arthropods, dominated by insects. Their time course of diversity, along with that of the other major Class of flying animals still extant, the birds, is shown in Fig. 4.6. The diversity of insects looks a lot like that of reptiles throughout the Paleozoic and Mesozoic, but more like that of mammals after the K-T extinction. Birds evolved from reptiles during the Jurassic, but, like mammals, showed little increase in diversity until after the flying reptiles disappeared at the K-T boundary.

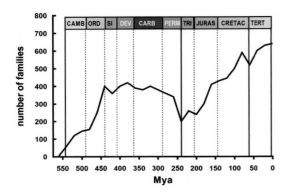

Fig. 4.3. Biodiversity of marine animals, measured by the number of families contained in the fossil record prior to the present in millions of years ago (Mya). The Cambrian, Ordovician, Silurian, Devonian, Carboniferous, and Permian Periods comprised the Paleozoic Era. The Triassic, Jurassic, and Cretaceous Periods made up the Mesozoic Era. The Cenozoic Era (the last 65 million years) consists of the Tertiary and Quaternary Periods. *Based on data from Signor (1994).*

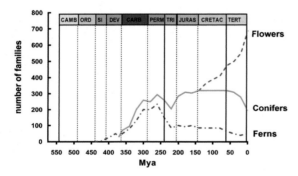

Fig. 4.4. Biodiversity of terrestrial ferns, conifers, and flowering plants over time. Geological ages as shown in Fig. 4.3. *Based on data from Signor (1994).*

Amphibians were the first vertebrates to diversify on land, beginning in the early Devonian (Fig. 4.5). Their slow acceleration in diversity was superseded by reptiles, which first appeared during the Carboniferous, prior to the crash at the P-T boundary. Reptiles recovered during the Triassic, but amphibians never did. Then reptiles began a long, steady decline in diversity with the appearance of the mammals, whose variety vastly accelerated following the K-T extinction that ended the dominance of the dinosaurs.

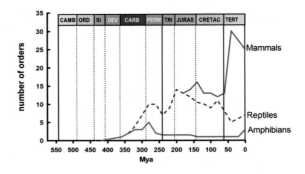

Fig. 4.5. Biodiversity of amphibians, reptiles, and mammals over time. Geological ages as defined in Fig. 4.3. *Based on data from Signor (1994).*

As this brief review reveals, biodiversity has shown a demonstrable net increase over time, but has been anything but steady and inexorable. Quite often, diversity has reached a steady state that is maintained for many millions of years, as for marine animals from the early Devonian to the late Permian (Fig. 4.3), and for reptiles throughout the Jurassic and Cretaceous (Fig. 4.5). Dramatic increases in diversity most often have occurred after ecological crises, perhaps precipitated by geophysical or meteorological catastrophes, leading to mass extinctions. Very often an increase in the diversity of one group supersedes the decline in another, as when conifers took over from ferns (Fig. 4.4), and mammals eclipsed reptiles (Fig. 4.5). Occasionally, however, diversity among different groups changes in tandem, as shown by insects and birds during the Tertiary (Fig. 4.6) after the flying reptiles were gone. In summary, while overall biodiversity does tend to increase over time, diversification within groups is episodic and dependent on ecological constraints, both among different groups of organisms, and between organisms and their environments.

Turning to the question of complexity, again the need for precise definitions is evident. As discussed in Chapter 1, complexity can be viewed in structural, functional, or algorithmic terms. Which definition fits best depends on the phenomenon under study. Since the origin of the universe, complexity has tended to increase primarily in structural (physical), then functional (biological), then algorithmic (social) forms, in that sequence. But every subject of study, from atomic structure to economic systems, has both structural and functional aspects, and at least a theoretical algorithmic definition. The aspect of complexity that is most evident for a given subject depends on the nature of that subject and the level at which it is analyzed. In this chapter on biological evolution, we will focus on Neubauer's straightforward use of the term complexity to mean the number of parts and the *number of functional relationships between those parts.*[39]

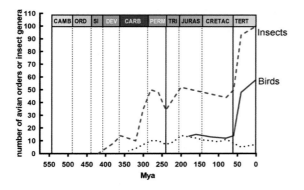

Fig. 4.6. Evolution of diversity in insects and birds, which evolved from reptiles (dashed black line). Geological ages as defined in Fig. 4.3. Based on data from Signor (1994).

If life began, as commonly assumed, in the form of simple solitary cells which diversified over time, it stands to reason that, as more forms evolved, some of them would logically have to have become more complex if life truly was in the simplest form possible in the beginning. A comparison of Earth's biosphere today with what it must have been like 3.5 billion years ago leaves no doubt that life has evolved into increasingly complex forms over time. But like diversity, empirical observations provide a nuanced picture of evolving complexity, and some unexpected observations.

Using growth in size as a simplistic proxy for structural complexity, the increase in organismic size over evolutionary time is readily evident in Fig. 4.7a. From a simple prokaryotic cell over 3 billion years ago to the largest organism extant today, size has increased by about 19 orders of magnitude. The vast majority of that increase, however, has occurred within the last one-third of the history of life. The effect is shown even more dramatically by plotting the number of different cell types—a better measure of complexity as defined above—which have evolved in the animal kingdom over the same

period (Fig. 4.7b). If data from plants and fungi were included, the cell type numbers would be smaller, but the overall trajectory and time course would be the same. The evolution of more than one cell type per individual organism has occurred almost entirely within the last one-sixth of the history of life.

Since proteins carry out the functions of the cell, the number of different proteins coded for genetically is one way to measure functional complexity. That plot (Fig. 4.8a) shows a progressive increase, especially among the metazoans, that corresponds to the onset of multicellularity about 600 million years ago. Note that the greatest number of protein-coding genes is seen in plants, with flowering plants being the most complex organisms evolved, by this criterion. Some of this can be explained by the tendency of plants to retain multiple copies of genetic information because of the way they reproduce, resulting in multiple possibilities for the mutational origin of new protein-coding genes. If they survive elimination by natural selection, they presumably have a function. This suggests that in sessile organisms, more diverse protein chemistry and metabolism may compensate for the organism's inability to move about in its environment.

In summary, complexity, like diversity, has shown a net increase over evolutionary time, but in nothing like a linear fashion. For two billion years, the level of complexity of living organisms remained basically static. Then, between a billion and 500 Mya, complexity began to increase dramatically. Because this coincides so closely with the rise of multicellularity, the increase in both complexity and diversity must be related to that epic transition in the history of life.

To what can the overall increase in organismic complexity be ascribed?

For those who view the need to explain at least the persistence, if not the origins, of complexity in terms of natural selection, the challenge remains to specify what the selective advantages of complexity are. We agree with those authors who argue in various ways that biodiversity inevitably leads to complexity, and that complexity has evolved over time to act as an increasingly effective adaptive system for resolving thermodynamic gradients.

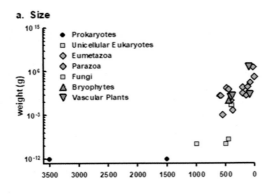

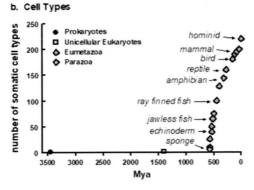

Fig. 4.7. Rise in structural complexity over evolutionary time, as measured by organismic size (a) and number of different somatic cell types (exclusive of neurons) per organism (b). The time point in million years ago (Mya) plotted for each organism coincides with the appearance of its ancestral form. Bryophytes are the mosses. Parazoa are the sponges. Metazoa are all the multicellular animals other than sponges. *Data on somatic cell number are from Valentine (1994).*

The thermodynamic argument goes thusly: The expanding universe leaves in its wake local inhomogeneities of matter and energy.[38] These result in gradients that cause the flow of energy from sources to sinks. The flow of energy through matter tends to organize that matter into adaptive systems for dissipating the energy gradients.[63] Stochastic processes, such as mutations of the genetic repository of organismic information and other emergent events like morphogenetic innovation[64] generate diversity, and that diversity inevitably leads to some adaptive systems that are more complex than others.[65] Those adaptive systems which can resolve energy gradients more effectively tend to be more complex,[66] and therefore enjoy a selective advantage. A more precise statement may be that adaptive systems evolve greater complexity to the extent that the increased complexity degrades more energy per unit of time

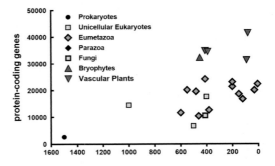

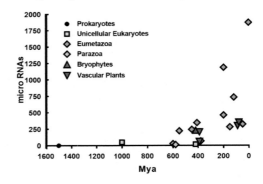

Fig. 4.8. Rise in functional complexity over evolutionary time, as measured by the number of protein coding genes per organism (a), and the number of regulatory genes as measured by microRNA transcripts per organism (b). The time point in million years ago (Mya) plotted for each organism coincides with the appearance of its ancestral type. Organisms were selected on the basis of data availability. Databases were accessed on 19 Nov 2013 for plant genes (http://plants.ensembl.org), animal genes (http://useast.ensembl.org), and microRNAs (www.mirbase.org).

and organismic mass (Fig. 4.9).[67] Either way, the thermodynamic point of view can be summarized by saying that living organisms are complex adaptive systems that have evolved to degrade energy.

At this point we encounter another one of those matters of perspective. Is the egg a chicken's way of making another chicken, or is the chicken an egg's way of making another egg? From the perspective of living organisms, natural selection is the means by which adaptations are perfected and new species are formed. From the perspective of the energy content of the universe, natural selection is the means for bringing about lasting changes that use energy and increase entropy more effectively. Both statements are factually defensible. While the origin of new species is a consequence of natural selection, to say that natural selection *happens in order* to make new species is to make a teleological assertion. But to say that natural selection *happens because* the PLA demands that energy be degraded and entropy increase in the most effective way possible, seems less so.

Great progress has thus been made in reconciling the evolution of life, which has tended to lead to increasingly complex organisms over time, with the SLT's imperative that free energy be degraded and entropy be increased overall in the organism and its environment. Notwithstanding the imperative of the SLT and the PLA, and the impressive way in which the authors cited above have succeeded in applying them to the evolutionary process, the actual time course over which diversity and complexity have increased in the living world does not lend itself to straightforward explanations. The universe has been expanding and entropy has been increasing steadily over the entire history of life on Earth. There would appear to be no facile way to explain why expansion and entropy alone could have so little effect for two billion years of relatively static and simple prokaryotic life, then generate one revolutionary innovation—the eukaryotic cell—about a billion and a half years ago, followed by an explosive increase in diversity and complexity only over the most recent 600 million years. It is tempting to suggest the action of some very significant threshold effects. Evolution of the eukaryotic cell was clearly one such threshold. The emergence of multicellularity was another. In both cases, structural complexity increased dramatically (Fig. 4.7), and with them a significant increase in functional and algorithmic complexity (Fig. 4.8) was seen.

Another threshold was reached when animals evolved with the ability to move about in their environments, seeking food and mates, including the pursuit of and flight from one another. This required the acquisition of detailed information about the environment in real time, and control of coordinated body movements to bring about locomotion. For that purpose—the management, analysis, and storage of large amounts of dynamic information—nervous systems evolved, and with them came behaving organisms, capable of altering

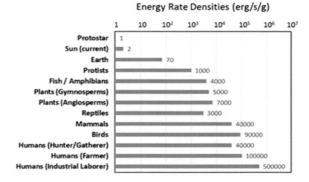

Fig. 4.9. Evolutionary increase in energy consumption by living organisms. Energy rate densities are rough estimates for an average entity in each category, based on data from Chaison, 2013. Entities are listed in their approximate order of evolutionary appearance.

their environments and interacting with one another. At this point, neural information became a supplement to genetic information as an added form of algorithmic complexity, with far reaching consequences for the evolution of life. That is the topic to which we turn next.

Summary of Chapter 4

The very definition of a living organism—as an enclosed, organized chemical system that draws energy from the environment to maintain its highly ordered state and carry out dynamic activity, including the ability to reproduce itself autonomously—suggests the ways in which biological change over time differs from changes in the purely physical world. Biological evolution is manifested on a much smaller scale, dependent on molecular interactions rather than the motion of large-scale physical forces, and, most significantly, characterized by a succession of entities that reproduce themselves by incorporating a repository of information that is passed on with high fidelity.

The origin of life is assumed to have arisen from a preexisting complex of minerals and prebiotic organic compounds that gradually, over multiple trial-and-error iterations, began to take on some of the characteristics of living cells. No one knows how this happened. All we do know is that eventually the process gave rise to a set of consistently recurring molecular interactions capable of consuming energy to maintain order. Enclosed by hydrophobic membranes that kept them sequestered from their aqueous surroundings, the protocellular packets of metabolic reactions achieved the ability to generate exact analogs of themselves, and to endow those analogs with sufficient information to recreate all the metabolic interactions as well as the means of their reassembly in successive generations. At this point, self-perpetuating cellular life came into being.

Prior to life's emergence, change had proceeded through the ponderous action of large-scale physical forces, with information playing no functional role. The appearance of living cells, as entities with finite lifespans capable of generating successive iterations of themselves, elevated the role of information. Periodically, that information could be altered by mutation (and after life discovered sex, by recombination). Those occasional changes modified metabolism and/or structure, usually in catastrophic ways that consigned them to oblivion, but on occasion in a manner that conferred an enhanced ability to survive and reproduce. By surviving in greater numbers through each reproductive iteration, the fortuitous changes became engrained in the informational repertoire of the organism's lineage. At this point, *natural selection* became a new force in evolution.

The relatively simple, single cell was a formula for success that persisted on Earth, presumably in many forms but without material alteration, for two billion years. Some cells (called autotrophs) early acquired the ability to draw energy from the sun for manufacturing their own organic nutrients from carbon dioxide, disgorging oxygen into the atmosphere as a bi-product. Others (the heterotrophs) subsisted by engulfing the autotrophs as their source of food. About a billion and a half years ago, some of the ingested autotrophs remained undigested inside the heterotrophs, conveying to the latter the benefit of various particular metabolic specializations. This excursion into symbiosis resulted in the eukaryotic cell, arguably the most revolutionary innovation in the history of life. About a half a billion years later, some eukaryotic cells ceased to separate after their reproductive divisions, and cellular clusters began to persist. With multiple cells interacting, some differentiation into specialized cell types was favored by natural selection, and that led to an explosion of myriad forms and functions —the first evidence of which begins to show up in the fossil record between 800 and 600 million years ago. The adoption of the multicellular life style was the next watershed event in the history of life. Soon, the multicellular forms evolved into separate lineages that today we recognize as plants (autotrophs), fungi (heterotrophs that digest their food externally), and animals (heterotrophs that digest their food internally).

References and Notes

[1] Irwin & Schulze-Makuch, 2011, 2020

[2] For a more elaborate discussion of this point, see Ch. 2 in Schulze-Makuch & Irwin, 2018

[3] A prevalent and not implausible theory about the origin of life on Earth is that it arrived here as cargo on a meteorite from outer space—perhaps from Mars or Venus or an even more remote location—and that all life on Earth is descended from that singular event. Even if so, life had to originate by an evolutionary process somewhere, and the original generative process wherever it occurred is the one with which we are here concerned. For simplicity, we will speak of it as though it occurred on Earth.

[4] Morowitz, 1968

[5] Oparin, 1938; Haldane, 1929; Orgel & Lohrmann, 1974

[6] Miller, 1953

[7] This argument is elaborated in Chapter 3 of Schulze-Makuch & Irwin, 2008

[8] Greenberg, 2000; Pizzarello, 2004; Chiavassa et al., 2005; Pizzarello & Huang, 2005; Sandford, 2008

[9] Cockell & Bland, 2005

[10] Matsuno & Swenson, 1999

[11] Hulshof & Ponnamperuma, 1976

[12] Fox & Dose, 1977; Lathe, 2004; Irwin & Schulze-Makuch, 2011

[13] Russell & Kanik, 2010

[14] Matsuno & Swenson, 1999; Neubauer, 2012. An alternative argument for the origin of life in continental hydrothermal springs has been made by Damer & Deamer, 2020.

[15] Muller, 1996

[16] Turian, 1999; Irwin & Schulze-Makuch, 2011; Neubauer, 2012

[17] Chaisson, 1987

[18] Orgel, 2003; Srivatsan, 2004

[19] Morowitz, 1968

[20] Orgel, 2000

21 Trefil, Morowitz & Smith, 2009

22 Norris, Loutelier-Bourhis & Thierry, 2012; Copley, Smith & Morowitz, 2007

[23] Cairns-Smith, 1982

[24] Fishkis, 2011

[25] Franchi & Gallori, 2005

[26] Norris, Loutelier-Bourhis & Thierry, 2012

[27] Russell & Kanik, 2010

[28] Elena & Lenski, 2003

[29] Taylor et al., 2009; Uroz et al., 2011

[30] Darwin, 1859

[31] Wright, 1932

[32] Gould, 1981

[33] Darwin, 1859

[34] Henderson, 1913

[35] Lotka, 1922

[36] Schroedinger, 1944

[37] Morowitz, 1968, 2002

[38] Chaisson, 1987, 2001

[39] Chaisson, 2001

[40] Neubauer, 2012

[41] Kaila & Annila, 2008

[42] Wurtz & Annila, 2010

[43] Sharma & Annila, 2007

[44] Kaila & Annila, 2008; Annila and Baverstock 2014. The authors point out that the difficulty in identifying genes closely linked to inherited traits and pathologies is that natural selection acts not on genetic variation per se, but on variation in any mechanism that alters the consumption of energy. This also explains why most disorders predictive of lifespan are disorders of metabolism.

[45] Gould, 1981; Eldredge, 1985

[46] Keosian, 1968; Cairns-Smith, 1982; Wächtershäuser, 1988; Lahav, 1994; Miyakawa et al., 2006

[47] Ergs are a measure of energy. When divided by units of time and mass, the dividend is a measure of energy flow density. All such measures in this section are taken from Chaisson (2001).

[48] Nicotinamide adenine dinucleotide phosphate (NADPH). This phosphorylated form of NADH is a high-energy compound important in light-driven biosynthesis.

[49] Cairns-Smith, 1982

[50] Alberts et al., 1989

[51] Schulze-Makuch & Irwin, 2008. An electron volt (eV) is another measure of energy, scaled better for extremely small units, like one photon or the reaction of one molecule.

[52] In actuality, electrons and protons from H are parceled out separately. Electrons pass from one electron-carrier to another, with each carrier lower in the chain having a stronger attraction for electrons than the carrier above it. Some of the energy released by this drop in potential

energy is used to pump the protons from H across a membrane, where their concentration rises to the point that they are driven down their concentration gradients by diffusion back through the membranes. The movement of protons through the membrane (called chemiosmosis) is what drives the formation of high-energy phosphate bonds in ATP. The end result is that the electrons and protons are eventually reunited when oxygen is reduced to water. We summarize these processes simply by saying that H moves down its gradient of potential energy.

[53] Campbell, 1996

[54] Purves et al., 1998

[55] Margulis, 2008

[56] Margulis & Sagan, 1995

[57] Schmidt-Nielsen, 1979

[58] Sharma & Annila, 2007

[59] Brooks & Wiley, 1988

[60] McShea & Brandon, 2010

[61] The Family is a taxonomic unit consisting of one or more genera, each of which contains one or more species. Crows, ravens, and jays belong to a single Family, the Corvidae. The Family Corvidae is one of many families in the Order Passeriformes, a higher, more inclusive unit.

[62] This is not to preclude the possibility that prebiotic organization may have included colonial aggregates of a variety of interacting protocells. That protocellular life was relatively undiversified and largely autonomous seems a reasonable assumption, however.

[63] Morowitz, 1968

[64] Goodwin, 1994

[65] Brooks & Wiley, 1988

[66] Gell-Mann, 1994

[67] Wicken, 1979; Chaisson, 2001; Neubauer, 2012

5
Evolution of Brain and Behavior

As living organisms became larger and more complex, they also became more dynamic in several ways. In developing from a single cell to a much larger multicellular organism, growth and development through a sequence of tissue and organ differentiations became part of the lifecycle. As parts of the organism became more modular, with different tissues and organs becoming specialized for specific functions, intercommunication and coordination among all the functional parts became a necessity. Growth and internal coordination are characteristics of all multicellular life, whether fungus, plant, or animal. Only animals, however, move about in their environments. Movements untethered to the substrate in which the organism thrives is the meaning of behavior as we will use it here. Evolution of behavior and the neural systems for processing information that control behavior are the subject of this chapter.

5.1 Behavioral Adjustments and Adaptations

From the earliest appearance of a living cell, information began to assume a greater role in directing the nature of change. With emergence of a genetic coding mechanism, the ability to perpetuate information made natural selection a new and novel force in the course of organic evolution—a factor absent from evolution in the inanimate world.

Behavior is inherently adaptive, since natural selection favors behaviors that are consistent with survival. As indicated in Chapter 4, selective survival and reproduction are distinctive features of biological evolution, and therefore critical to the contribution of natural selection to energy dissipation as dictated by the Second Law of Thermodynamics (SLT) and the Principle of Least Action (PLA). Reproductive behavior ensures the perpetuation of the species. Avoidance behavior diverts the organism from danger. Prey seeking and foraging behavior acquire food (potential energy). Migration enables relocation to favorable environments, seasonally or as needed. Social behavior is geared toward collective actions that benefit the group as a whole.

Every behavioral act has potential consequences. Survival is obviously promoted by mechanisms that impress those consequences upon the animal—enabling it to repeat and profit from behavior with favorable outcomes, while avoiding those behaviors that lead to unfavorable consequences. Selective reinforcement of this type assumes a place within the life of an animal that natural selection plays over the life history of the species, promoting survival and augmenting fitness. Implicit in such a capability is the capacity to acquire new information from the environment and store it for at least a period of time within the life of the individual animal.

When behavior emerged as a characteristic of animals, therefore, a new form of information processing and storage beyond what genetic coding could provide became necessary. This new form of information processing had to deal with information gleaned from the environment, within the life span of the animal in real time, even if never previously encountered. It had to be updatable from moment to moment, in a form appropriate for integration into the animal's current biological state, and capable of instigating appropriate adjustments and responses. Two different but interrelated systems, using chemical and neural signaling, respectively, evolved to enable the animal to engage with its environment.

5.2 Origins of Chemical Information Signaling

Sensitivity and reaction to the environment is, of course, not unique to animals. The earliest cells had to have some means of responding to favorable nutrients and physical conditions in their environments and avoiding harmful substances and circumstances. Sensitivity to specific chemical and physical stimuli has therefore been an inherent capability of living cells from the beginning, of necessity. Reaction in the form of chemical signaling has correspondingly been an ancient process of living cells.

Autocrine signaling is probably the oldest form of communication for living cells, as it serves to coordinate the activity of a cell in response to its own metabolic products and secretions. As multicellular organisms evolved, a mechanism known as **paracrine** signaling became necessary for communication between different cells a small distance apart. In metazoa—animals with three tissue layers and an interstitial space between layers—substances are emitted into the space between cells, which provide an avenue for diffusion to nearby target cells. The further evolution of metazoans with well-differentiated tissues and organs interconnected through a circulatory system provided a mechanism for chemical signaling to all parts of an organism, by secretions into the circulatory system for transport to distant target cells. Organs that generated and released substances for communication by this mode are known as **endocrine** organs, and their secretory products are **hormones**.

Most hormones fall into one of two broad categories: (1) steroids or similar lipid-soluble molecules like prostaglandins, or (2) amino acid derivatives, peptides, and proteins. All metazoans secrete both categories of hormones.

Steroid hormones are secreted primarily from gonads and other internal organs, and are involved primarily in reproductive coordination and metabolic regulation. Peptide and protein hormones also regulate metabolic functions, and are involved especially in the regulation of growth and maturation.

The earliest nerve cells made use of another category of chemical signals referred to as neurohumors. These are small molecules, either actual amino acids or their derivates and close chemical analogs. Most such chemical signaling occurs across the tiny space between two nerve cells, in which case the substance is known as a **neurotransmitter**. However, nerve cells also secrete other molecules for transport through the blood stream for action at a distance, which qualifies them as hormones. When their cell of origin is a nerve cell, they are called **neurohormones**.

5.3 Origins of Neural Information Processing

In the course of evolution, all cells evolved with a potential energy differential across their bounding membranes. By selectively pumping out some ions and preferentially admitting others into the cell's interior, concentration gradients developed across the membrane, leaving it electrically unbalanced, or polarized. Physical and chemical stimuli of relevance to the cell acquired the ability to alter the membrane's permeability to certain of these ions—mainly, potassium (K+), sodium (Na+), and chloride (Cl-)—depending on the cell. Because these ions carry an electrical charge, their movement across the membrane constituted a flow of current. Most often, the current restored the electrical balance, or "depolarized" the membrane; but occasionally the polarity was increased and the membrane became "hyperpolarized." Just as certain stimuli could elicit either depolarization or hyperpolarization, so could the current fluxes generate cellular responses, such as movement toward or away from the precipitating stimulus.

As multicellular organisms evolved—and particularly as animals with their need to move about emerged—some cells evolved with a heightened ability to respond to specific environmental conditions by becoming depolarized or hyperpolarized, and to transmit that information to other parts of the organism. At the same time, development of a motor apparatus—muscles for moving the organism, and other effectors such as glands for generating chemical signals, and defensive responses—came under the coordinated control of the whole animal. Thus, the need for a bidirectional flow of information arose, in the form of travelling waves of depolarization or hyperpolarization—from the exterior into the animal, to inform the animal about its environment, and from the interior to appropriate effectors (muscles or glands), in order for the animal to respond appropriately.

Cells which we now call neurons became specialized for this long-distance transfer of information. Thin fibers (axons) grew out from the cell's nucleus-containing main body (soma) to reach points at a distance ranging from micrometers to meters. Waves of (usually) depolarizing current fluxes could thus travel along the elongated axons, quickly conveying information over a considerable distance. When the travelling wave of depolarization reached the end of the neuron's extension, it contacted another cell at a junction—the synapse—and transmitted the information to the adjacent cell by releasing a chemical, or neurotransmitter. In some cases, the information could be transmitted purely by electrical excitation, though chemical transmission is now by far the more common and was almost certainly the ancestral condition.

Over long distances, the travelling wave had to be self-sustaining (which the inherent potential energy gradient all along the polarized membrane made possible), like a moving wave of falling dominoes. Conduction of a succession of depolarizing waves thus delivered information in all-or-none, or digital, bits. Locally, waves of either depolarization or hyperpolarization could alter the cell's response—by releasing neurotransmitters or other secretory products in proportion to the degree of the polarity change. This type of information was thus conveyed in an analog form. Thus neurons became specialized for conveying information over short or long distances that enabled the organism to respond to its environment in an appropriate manner and quickly enough to have survival value.

5.4 Evolution of Invertebrate Nervous Systems

5.4.1 Nerve Nets

The simplest nervous systems are nets of neurons providing pathways for transferring information from one point to all other points of a radial organism (Fig. 5.1a). A lattice-work of branching connections can thus connect any point on the organism's surface to any other point in the organism's body. Stimuli could come from any direction, and the appropriate response could also be in any direction. A dispersed network of neural connections, coordinating the organism's behavior in a straightforward stimulus-response manner, serves the needs of radial animals like jellyfish and sea anemones quite well. Such a network efficiently coordinates the simple behavior these animals require for capturing and consuming their food, consistent with the PLA.

5.4.2 Nervous Systems of Flatworms

The more ancient phyla of worms—the flatworms and round worms—have relatively simple behavioral requirements: the need to approach favorable environments and retreat from deleterious situations. They are able to rely mostly on simple stimulus-response behaviors. The ability to respond to stimuli in a coordinated, adaptive fashion required the ability to get information from the environment to the appropriate effector cells. In these ancient phyla, then, are found the first true central nervous systems.

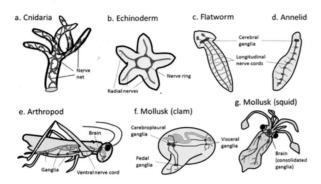

Fig. 5.1. Diversity of invertebrate nervous systems. (a) Nerve net; (b) radial nerve organizations; (c) bilateral nerve cords; (d) fused nerve cord with minimal bilateral brain; (e) consolidated ganglia and brain; and (f) dispersed ganglia with a complex brain. *Art by Louis Irwin, adapted from Campbell & Reece, Biology, 8th Ed. ©Pearson Education, Inc.*

Even in the most ancient of longitudinal body plans, the directionality of forward movement meant that the densest input and most relevant information came from the front of the animal—from the direction toward which it was heading. This meant that sensory organs were located mainly in the forward (anterior) portions of the animal, and whatever integration and coordination took place could best be handled in this forward location. As a greater concentration of neurons handling more information became located anteriorly, this region of nervous systems became enlarged, and over time, evolved to exert more front-to-back hierarchical control. Thus brains came into existence.

With the innovation of bilateral body plans accompanied by a greater degree of directional (mostly forward) motility, transmission of information over the animal's longitudinal axis became important. Some transmission of information from one side of the animal to another was still needed, and short, local circuits originating in neural cell soma at the periphery, served this purpose adequately. As the body plans became longer, however, information needed to be transmitted up and down the length of the animal. Axonal extensions thus became clustered in parallel bundles running lengthwise along the animal's long axis. Ancestrally, these nerve cords were paired, with two nerve cords running parallel along the animal's ventral (bottom) surface. Communication between the two cords was carried out through perpendicular connections (commissures), which tended to become bundled, giving the overall nervous system a ladder-like appearance along the animal's ventral aspect (Fig. 5.1c).

5.4.3 Nervous Systems of Annelids and Arthropods

The Annelids, to which earthworms belong, have segmented bodies, in which each segment along the length of the worm's body, acts semi-autonomously. The muscles in each segment can constrict or relax, and by doing so in a concerted, rhythmic fashion from front-to-back, can move the animal forward through the soil. Each segment is controlled by a concentration of neurons, referred to as ganglia, though overall coordination is achieved by regulating the timing with which each segment is activated to contract. Much of the coordination appears to be achieved by local pacemaker-like interactions, rather than by a high degree of central coordination.

At least six sensory modalities have been documented for Annelids. They can detect and respond at least to the following: touch, pressure, vibration, light, shadow, and irritation (pain?). The integration of this sensory information, and the generation of some motor coordination, occurs along the length of the animal, but quantitatively more information is processed at the anterior end of the nerve cord. Furthermore, the mouth takes in food at the animal's anterior end, so some neural coordination of oral operations is locally required.

Therefore, the nervous system is somewhat expanded and differentiated into several ganglia connected by commissures, or (in some forms) fused into a single ring of neural tissue surrounding the mouth and pharyngeal end of the digestive tract (Fig. 5.1d). This expansion and elaboration of neural tissue constitutes the brain.

Arthropods, like crustaceans, insects, and arachnids, are descended from annelid-like ancestors (with insects retaining the annelid body plan in the larval stages of their life cycles). Their nervous systems accordingly retain some ancient architectural features, including ganglia concentrated in specific segments, a ventral nerve cord, and fused ganglia constituting an annular brain surrounding the pharyngeal region of the head (Fig. 5.1e). The degree of sensory input and integration, as well as motor control, is much more elaborate in most Arthropods than in Annelids, but structurally, their nervous systems are primarily variations on a theme.

5.4.4 Nervous Systems of Mollusks

No phylum consists of a greater range of complexity than that of the Mollusca. At the simplest end are the bivalves. These are sessile animals enclosed in a mineralized shell, like clams and oysters. They are filter-feeders that remain motionless for their entire life cycle except for opening and closing their shells. Their behavior is minimal, their need for coordination almost entirely a matter of internal physiological adjustments. At the opposite extreme are the cephalopods, like the squid and octopus. These are active predators with long, highly manipulable arms and a complex behavioral repertoire. They have a high degree of manual dexterity, an exquisite tactile sense, superior vision, and clear problem-solving abilities. They expend more energy in the process of feeding higher up the food chain, thereby consuming and degrading more energy than their less active predecessors.

Ancestrally, mollusks consisted of animals with an amorphous body plan, anchored by a "foot" which provided a means of motility. Consistent with the scattered organization of its internal organs, the nervous system ancestrally consisted of a set of six paired neuronal concentrations, or ganglia, each controlling a particular bodily function (Fig. 5.1f).

These ganglia communicated with one another through commissures that tied them together into a unified, essentially circular, nervous system. As more complex body plans and behavioral repertoires emerged within the phylum, different ganglia became enlarged, and tended to fuse together. This process led to brains in cephalopods as large and complex as that of some fishes (Fig. 5.1g), while those forms that remained simple and with sessile life styles retained nervous systems barely distinguishable from those of flatworms.[1]

It should be noted that the neural complexity of the cephalopods is not a recent phenomenon. Predatory nautiloids, presumably with already well-developed nervous systems, were swimming in Ordovician seas 500 million years ago, when the simplest vertebrates were in their evolutionary infancy. At the same time, simple gastropods like clams with the least complex nervous systems in the phylum, persist to the present day.

The overall picture of neural evolution in the invertebrates is one of increasing complexity over time. But that is only on average. Within a given phylum, conservative simplicity is more often than not the rule, though changes across phyla do show the net trend toward expanded size and complexity. Flatworms have more complex and centralized nervous systems than sponges. Annelids and arthropods show even more centralization and cephalization (forward concentration) of neural control. Complexity has reached a conspicuous level among the Cephalopods, though most other mollusks have retained much simpler nervous systems. From the simplest to the most complex, each degree of neural organization has evolved to provide an optimal level of coordination for the extent of energy capture and consumption that each form requires.

5.5 Evolution of Vertebrate Nervous Systems

Though constituting but another subphylum of the animal kingdom, we here treat vertebrates in a section of their own, because of certain qualitative differences between vertebrates (animals with backbones) and all the invertebrates. The first of these is that neurons take on a more complex morphology, with bipolar and multipolar neurons becoming much more common. This means that information from other neurons can impinge on their target neurons along a fibrous extension (**dendrite**) which conveys the excitation to the target's soma, rather than just onto the cell body directly. This appears to markedly increase the degree to which any given neuron is connected by many others. Secondly, the ganglia of invertebrates tend to be composed of a fibrous mat of axons surrounded by a rind of cell soma. In vertebrates, neuronal cell soma tend to be scattered, often in layer-like fashion, among the coursing fibers of axons and dendrites. This appears to minimize construction costs and fiber lengths, both of which are energetically costly, among the denser populations of neurons in vertebrates. This innovation is a direct expression of the demands of the PLA for the optimal use of energy toward a functional end. Thirdly, the nerve cord in vertebrates is always single, dorsally situated, and centrally located, rather than being a paired or fused, ventrally situated structure. Finally, brains are almost universally larger in proportion to body size among the vertebrates than they are among the invertebrates.

Because vertebrates have been studied so extensively, a vast data set is available for analyzing the relationship between brain size and body mass. That information shows a consistent and statistically robust allometric relationship between brain and body size, meaning that brain size and body size vary linearly when plotted on log-log scales (Fig. 5.2). Across all vertebrates, the slope of the line approximates 2/3, though slight variation does occur within taxonomic groups. This suggests that underlying whatever specific influences may affect the evolution of particular brains, there are inherent constraints on brain size related to the size of the animal in which the brain operates. A fractional exponent like 2/3 mirrors the relationship between surface (linear dimension squared) and volume (linear dimension cubed), though this in turn could reflect a number of different physiological and biomechanical factors. The relationship is highly

consistent within taxonomic groups, but species cluster separately across certain Classes. Generally speaking, brains of birds and mammals plot higher in relation to body size than do the brains of reptiles, amphibians, and fishes.[2]

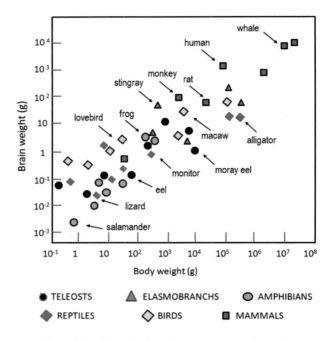

Fig. 5.2. Brain-body mass ratios in vertebrates. Brains increase in proportion to body mass with a slope approximating 2/3 when plotted on log-log scales.[2] *Redrawn from data cited in Striedter (2005).*

5.5.1 Brain Evolution in Fishes

In popular usage, the term "fish" can refer to any completely aquatic vertebrate that extracts dissolved oxygen by passing water over gills. This covers a broad range of taxonomic groups, collectively accounting for more species than all other vertebrates combined.

The ancestors of all vertebrates were jawless fishes, of which only a few species remain. The archetype of the vertebrate brain, which first appeared in ancestral jawless fishes, has been conserved throughout the whole of vertebrate evolution.[3] It consists of three major divisions (Fig. 5.3). First is the hindbrain (rhombencephalon), concerned primarily with coordinating stereotypical motor and visceral functions, like breathing, swallowing, and coughing, as well as sensory inputs related to balance, vibrations, and electrical fields. It is an expansion at the immediate anterior end of the spinal cord. Secondly, the midbrain (mesencephalon) lies in front of the hindbrain. Ancestrally, its primary function was the processing of visual inputs. Thirdly, the forebrain (prosencephalon) lies in the most anterior position. It consists of two prominent regions—the diencephalon, concerned largely with autonomous physiological regulation but also serving as a major relay center, and the telencephalon, which ancestrally appears to have been concerned mainly with processing olfactory information. Even in their earliest forms, all vertebrates showed this significant degree of encephalization.

The smallest brains among the vertebrates are found in the jawless fishes (Fig. 5.4). Lampreys, which are parasitic with little need for active behavioral regulation, have brains as small as 30 mg in a body weighing up to 55 grams.[4] Their midbrain receives sensory inputs that are topologically organized, meaning that specific neurons are arrayed spatially in a way that parallels the spatial configuration of the environment from which the stimuli come. The topological orientation of sensory fields is therefore likely to have been an ancestral principle of neural organization. The cerebellum—an elaboration of the hindbrain particularly important for sensorimotor coordination—is hardly more than a narrow band of tissue in these animals.

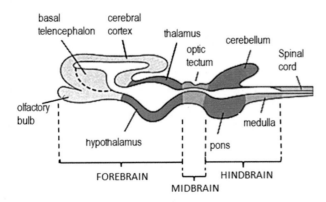

Fig. 5.3. The generic vertebrate brain. *Adapted from graphic design by Denis Paquet, in "The Brain from Top to Bottom" (https://thebrain.mcgill.ca/), with permission.*

The jawed fishes are today regarded as having diverged into three distinct directions: one toward the cartilaginous fishes (Chondrichthyans), a second toward the ray-finned fishes (Actinopterygians), the third toward the lobe-finned fishes and their tetrapod successors (Sarcopterygians).[5] The brains of all the jawed fishes range from somewhat larger to much larger than in the jawless fishes, and are highly variable in complexity.

Among the bony fishes, the radiation leading to the ray-finned fishes gave rise to more species, in more variable forms, than all other vertebrate groups combined. This diversity is reflected in a large range of brain sizes, specializations, and complexities.[6] Bottom feeding fish have large medullary regions, where facial nerves innervate the brain, probably reflecting the role of taste receptors for discriminating between edible and inedible food. Fish that rely on olfaction have larger forebrains, which process olfactory information. Those fishes that are specialized for generating and/or sensing electrical currents in their surroundings often have unusually large cerebellar enlargements extending from their hindbrains.

The radiation that led to the lobe-finned fishes also gave rise to all land-dwelling vertebrates (tetrapods). The living marine survivors of this lineage have generalized brain structures showing no particular specializations, more akin to what was likely the ancestral vertebrate pattern than the more elaborated variations seen among modern ray-finned fishes. Since they are few in number, however, we can't be certain how variable brains may have been in ancestral lobe-finned fishes.

Brain size and complexity also varies among the cartilaginous fishes (the sharks, skates, and rays), ranging from a small brain with minimal cerebellum and forebrain in the megamouth shark, to a much larger brain with significantly enlarged forebrain and cerebellum in the manta ray. As a group, the cartilaginous fishes have brain-to-body mass ratios that fall above that of the ray-finned fishes, amphibians, and reptiles, and a little below that of birds and mammals.[7]

5.5.2 Brain Evolution in Amphibians and Reptiles

The evolution of amphibians from lobe-finned fishes enabled access to a huge increase in previously untapped forms of energy in keeping with the PLA.

However this land invasion was not accompanied by an immediate dramatic increase in brain size or complexity, or in energy utilization (Fig. 4.9). The main change appeared in the midbrain, where projection of visual information acquired greater importance (perhaps because light is transmitted more clearly through air than through water) in frogs and toads. Among the salamanders, for whom vision is generally less important, overall brain size and complexity has actually diminished over evolutionary time.

Evolution of reptiles from amphibians was a major developmental and physiological achievement, but it appears to have required no more increase in brain size than would be predicted from body mass. The regression plots for brain size in reptiles fall well within the range for those of amphibians and the ray-finned fishes.[8]

In all tetrapods, the insertion of a new dorsal pallium between the preexisting medial and lateral pallia of the telencephalon can be discerned.[9] This region receives a richer and more varied input of sensory information than is seen in the strictly aquatic vertebrates. One change that signaled the expansion of the telencephalon that was to come in mammals, was the beginning of an expansion along the dorsal midline in turtles.[10]

5.5.3 Brain Evolution in Mammals and Birds

Mammals evolved from reptiles during the early Mesozoic, over 200 million years ago. Birds evolved independently from a separate reptilian line some 30 million years later. Both groups are characterized by having significantly larger brain-to-body mass ratios than do all other vertebrates. And in both groups, a disproportionate increase in the telencephalon, and a significant increase in the cerebellum, account largely for the increase in brain size. This relative increase in brain size began in the ancestral forms of both the avian and mammalian lineages.[11]

Despite their early appearance, these relatively larger brains did not enable their possessors to replace the relatively smaller brained dinosaurs for 100 million years, when global extinctions opened up niches to which the dinosaurs had been well enough adapted despite their smaller brains.

While birds and mammals both showed dramatic increases in forebrain size and complexity from early stages in their divergent evolutionary histories, the nature of the changes were significantly different. In mammals, the dorsal pallium became greatly enlarged, eventually enveloping the entire dorsal aspect of the brain. This expansion is referred to as the neopallium. In birds, the vast enlargement came from the striatum, a different ancestral region of the forebrain, leading to the neostriatum. It too, eventually became the dominant feature of the entire bird brain's structure. Another difference is that the optic tectum, an outgrowth of the midbrain, became much more massive in birds, while in mammals this region shrank in relative size, as optic information came to be relayed into the newly massive neopallium. The terminology here can get technical. The major point is that brain enlargement, characterized mainly by enlargement of the forebrain, occurred to a larger degree in mammals and birds than it had in earlier vertebrates, but it did so in different ways. An expanded forebrain in mammals and birds was therefore a case of parallel evolution—two distinctive ways of adapting to the demands of avian and mammalian lifestyles.[12] The obvious question is, what were the selective forces driving those changes?

At the dawn of the mammals, the daytime niches were filled with ruling reptiles. The nights were safer for the smaller vulnerable mammals, but were cooler and devoid of light. Mammals thus needed to stabilize their body temperatures at a higher level (homeothermy) in order to remain mobile, and had to depend on non-visual senses, especially olfaction and hearing, for information about their environments. Unlike vision, which conveys spatially explicit information about the external world, olfaction and hearing provide qualitative information requiring memory and its related faculty, a good sense of temporal ordering, in order to be effective. While birds, occupying mainly daytime niches, continued to rely heavily on vision, their dynamic activity in three-dimensional space required a higher degree of coordination between high-definition visual input and the real-time

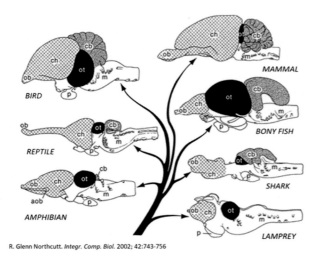

R. Glenn Northcutt. *Integr. Comp. Biol.* 2002; 42:743-756

Fig. 5.4. Evolution of the brain in vertebrates. Shown are samples of brains from different Classes of vertebrates. Within each Class, there is a great deal of variation, and no brain should be considered typical. Key: aob, accessory olfactory bulb (cross-hatched); cb, cerebellum (stippled); ch, cerebral hemispheres (cross-hatched); m, medulla oblongata; ob, olfactory bulb (cross-hatched); ot, optic tectum (black); and p, pituitary gland. *Credit: Northcutt (2002), Oxford Journals, with permission.*

sensorimotor adjustments that flight demanded. It is this need to integrate heightened sensitivity to new sensory modalities, along with the added sensorimotor apparatus required to maintain homeothermy that is thought to have placed a premium on the expansion of the telencephalon in mammals and birds. Taken together these nervous system innovations in birds and mammals compared to reptiles relate directly to their significant increase in energy utilization.

5.5.4 Brain Evolution in Hominins

Humans do not have the largest brains of any animal—elephants and whales own that distinction—but they have the largest brain-to-body mass ratios by far. Primates as a group have larger brains than would be predicted by their body sizes; but within the primates, the hominin line leading to modern humans began to exceed in relative brain size that of the other great apes when australopithecines evolved about four million years ago.[13] This was followed by two dramatic increments in relative brain size. The first occurred a little over

two million years ago when *Homo* split off from its ancestral genus, *Australopithecus*; the second came when modern *Homo sapiens* emerged about 100,000 years ago.[14] For the two million years between *Homo habilis* and the extinction of *Homo erectus*, relative brain size increased very little, and has remained in a plateau since the final burst in relative size (largely attributable to a decrease in body size) over the last 100,000 years. Like that of mammals in general, most of the size increase in the brain is attributable to enlargement of the neopallium.

The human brain does not differ in gross anatomy from that of the chimpanzee, its closest living relative. A few organizational features do appear to be unique to humans, however. These include expansion of the dorsal temporal lobe, which processes auditory information related to speech, and expansion of regions of the parietal lobe, which include somatosensory cortex. In addition, the human neocortex makes more robust and direct connections between motor control areas and all the major vocal motor control areas in the medulla and spinal cord. Finally, the lateral prefrontal cortex (though not the entire frontal lobe) appears to have become enlarged as the genus *Homo* evolved.[15] In summary, the human brain has become both larger, and, through organizational innovations, more complex, over evolutionary time.

5.5.5 Summary of Trends in Brain Evolution

The net tendency has been for brains to become larger and more complex on average over the course of evolution. When brain enlargement and complexity have occurred, they have tended to emerge early in the history of a lineage, and more often than not are restricted to a minority of the members of a given taxonomic group. Among the invertebrates, gradations in neural complexity are seen within the same phylum. Earthworms have simple nerve cords with only small anterior enlargements for brains, while some polychaete worms in the same phylum have brains rivaling the complexity of some mollusks. Among the latter, most Classes have relatively simple, dispersed neural ganglia, but cephalopods have brains as large and complex as that of some fishes. Within the same

Arthropod Class, the Crustacea, shrimp have simple ladder-like nerve cords with an annular brain barely distinguishable from any other ganglion, while the nervous system of crabs is largely a central consolidation of ganglia.[16] Among vertebrates, brain enlargement beyond what would be projected for given body sizes has been restricted to relatively few groups, like the psittaciformes (parrots, macaws, and cockatoos) and corvids (crows, ravens, and jays) in birds, and cetaceans (dolphins and whales), elephants, and primates among the mammals. Increasing size and complexity, however, are not inevitable; regression in both size and complexity has occurred in some groups, as in lungfish and salamanders. All these cases appear to reflect parallel and independent evolutionary changes. Relative brain size and complexity have increased most spectacularly in humans, but this is a recent event compared to similar changes that appeared independently about 20 million years earlier in psittaciformes and cetaceans, not to mention over 400 million years ago in cephalopods. [17]

5.6 Evolution of Increased Cognitive Capacity

When nervous systems appeared in the course of evolution, they enabled organisms to become engaged with their environments in ever more diverse and complex ways. To increasing degrees, animals were freed from the immediate impact of their surroundings. Sensory input gave them an internal connection with their external world. Motor coordination gave them the capacity to resist the forces impinging upon them, generate new ones to the point of changing their environment, or at least their situation within it. Thus animals were able to become increasingly and proactively independent actors in the ecosystems of which they were a part.

To the extent that disengagement from the environment was a positive adaptation, it was restricted by natural selection in ways that favored survival. Thus, not all sensory input that the environment could provide was relevant to a given animal in a particular situation. Color vision was of no use to a nocturnal animal, so neural architecture otherwise devoted to that would be better spent on a high degree of olfactory sensitivity. Not all motor capabilities were equally useful; speed was more important than strength for small prey, while stealth and strength better served the predator. As natural selection honed the sensory abilities and motor capacities of each species according to the niche each occupied, patterns of information processing and control were laid down in the nervous system of each that molded the animal's ability to navigate through its daily life. Those patterns of information processing by definition were species-specific.

The argument that the capacity for behavioral complexity and flexibility—what is popularly meant in a confounded sense most often by the term "intelligence"—has evolved over the course of evolution, is an argument at once easy to make and deceptive. It is easy to make in the sense that some nervous systems are more elaborate than others, some brains are larger than others, behavior is more complex in some animals than in others, and the trend over time has been a net increase in all those features. The presumption that all these cases represent a greater overall capacity for information processing is logically defensible if not intuitively obvious, and examples of increased capacity have clearly emerged over evolutionary time.

The case for a linear increase in behavioral flexibility over time is harder to make, for two reasons. First, the evolution of truly insightful and flexible behavior has occurred less often than logic would lead us to expect, based on the presumption that greater insight and flexibility are inevitably a selective advantage. What is easier to defend is the proposition that "just enough" flexibility, at just the right level of complexity, has evolved to promote the survival of each species. It is the adaptability of actions or behaviors maximally appropriate to each species, enabled by specific patterns of information processing, that appears to be the target of natural selection rather than complexity of information processed in a general sense.[18] Secondly, when conspicuous cognitive capacity has evolved, it has done so independently and non-linearly. The laminar cortical architecture characteristic of high-information processing brain regions appeared first in an invertebrate, the cephalopods (squid and octopi), over 400 million years ago. Entirely independently, certain birds—the psittacines like parrots, and corvids like crows—probably by 20 million years ago had developed the capacity for complex anticipatory behavior. Again, entirely independently, some marine mammals had achieved a brain size larger than that of modern humans by about the same time, around 20 million years ago. The brains of protohumans began to exceed that of chimpanzees only 4 million years ago, and by an independent trajectory from that of marine mammals.

There has been an unquestionably explosive growth in the human brain since the hominid divergence from other primates. Since most of the growth has been attributed to expansion of the forebrain, where insight and anticipatory analysis are centered, there has been a tendency to view anticipatory, analytical, problem-solving behavior as an end-point favored by natural selection. While certainly one measure of cognitive advancement, this definition is clearly self-referential. The cognitive abilities of other animals with large and complex brains are largely unknown to us. The remarkable sensory-motor manipulative ability of cephalopods, the long-range communication skills and social habits of the cetaceans, and exquisite auditory sense and social memory of elephants, extend to realms of which we have little understanding. Those capacities evolved over time through changes in the brains and nervous systems that have been unique to each of those groups. At the same time, most members of the animal kingdom have persisted at a plateau of neural complexity totally adequate for their needs, once the defining behaviors and nervous systems subserving them first emerged.

Notwithstanding the above, is there a broader argument that can be made about the reason that, on average, brains, nervous systems, and information processing capacity have increased in complexity over time? We expect that a careful analysis will disclose that all behaviors and the neural infrastructure that enable them, however simple or complex, serve to maximize the efficiency or effectiveness with which the organism makes use of energy at its particular node in the ecosystem. This is supported by the fact that evolution has given rise to organisms with increasing energy rate densities over time (see Fig. 4.9).[19] Thus, we are brought once again to the fact that the arc of evolutionary history has followed the mandate of the Principle of Least Action, degrading energy ever more efficiently and effectively in the direction required by the Second Law.

5.7 Energy and Evolution of the Human Brain

Energetic considerations surely played a major role in brain enlargement of our protohuman ancestors. Given that the human brain requires an inordinate proportion of the body's available energy, it is important that it has evolved to perform efficiently. The neural networks of the human brain have evolved to yield maximum communication effectiveness while minimizing connection cost (maximizing efficiency).[20] Both effectiveness and efficiency have been selected for. The importance of this analysis is shown by the fact that degradation in effectiveness of this optimally functioning connectivity is a characteristic of autism and an early predictor of schizophrenic pathology.[21]

If organisms evolved to transfer energy effectively, there should be evidence of a planning mechanism capable of cost-benefit analysis of prospective outcomes. Brain structures associated with reward and planning have been identified as signaling the relative cost and benefits of anticipated effort and ensuing reward in humans.[22] More specifically, a neural measure of appreciation which predicts future consumption is increased activation of the medial prefrontal cortex and dopaminergic reward circuits of the brain while the respondent anticipates and receives reward during a performance. Activation of the dopaminergic circuits results in the increased utilization of brain glucose. Brain structures have been identified that assess the cost of effort against reward.[23] The release of dopamine by the brain signals cost-benefit analysis, and brain functions have been related to the individual's analysis of the optimality of planned actions.[24] Some mental pathologies increase energy intake and output for personal gain beyond the socially accepted in-group norms, as seen in addiction, manic behavior and psychopathy. Congruently, these states are often experienced as not unpleasant, and show elevated levels of blood glucose in brain reward structures, such as dopamine mediated effects in the nucleus accumbens.[25] Chronic activation of the reward pathways (primarily dopaminergic and opioid), as in an indulgence in caloric high energy foods or recreational drugs, leads to the enhanced response of these structures to addictive cues, chronic craving, and overconsumption.[26]

It is not surprising that a maternal diet of energy rich foods activates these circuits and induces preferences for such foods in the young.[27] These energetic considerations emphasize the unity of humanity, not only with other life forms but also with the basic processes of the inanimate physical and chemical world; yet conversely, and congruently, the model identifies the uniqueness of the human species. We are the supreme energy dissipaters. This is clearly manifest in our species' greater effectiveness in transforming energy, not only through exaggerated fuel consumption, but especially in the emergent manifestation of energy transduction—namely the transformation of information, through its creation, storage and exchange. This is the topic to which we turn next.

Summary of Chapter 5

All cells are sensitive to their physical and chemical environments, thus are capable of reacting to chemical signals. As multicellular organisms arose, the ancient capability of chemical signaling became a means of integrating the activity of cells and organs throughout the organism. Hormones evolved as messengers secreted by endocrine glands for distribution through the circulatory system to distant targets. The process of secreting neurohumors into the extracellular space for signaling over short distances became a way of communication among certain cells specialized for transmission of information rapidly to specific bodily locations. These cells, called neurons, became elongated and specialized for conducting waves of membrane excitation or inhibition from one end of the cell to another. They activated or inhibited other nerve cells, or effectors like muscles or endocrine glands, by releasing specific neurohumors called neurotransmitters. The overall collection of neurons became more numerous and complexly interconnected, forming nervous systems.

With the evolution of nervous systems, information took a quantum leap forward in governing the trajectory of animal evolution. In its architecture and function, the nervous system is the most complex organ of an animal, and—not incidentally—the most energetically demanding per unit of mass. The end result of the evolution of nervous systems has been the capacity for animals to become active participants in the web of life, beyond the passive role played by the other kingdoms that occupy the biosphere. Natural selection has enshrined nervous systems as integral components in the makeup of animals, since they promote survival by enabling an animal to react to and operate within its environment in a more advantageous way.

Inasmuch as animals in general have evolved into larger and more complex forms over time, so too, on average, have the nervous systems that serve to integrate sensorimotor information, regulate their physiology, and control their behavior. Conspicuous advances in neural complexity are seen in the evolution of flatworms, then annelids and arthropods in one direction and mollusks in the other. Within each of these groups, a few have developed much more complex and integrated nervous systems than others, while the majority have remained similar to that of their simpler ancestors. Among the mollusks, the most spectacular increment in size and complexity is seen in the cephalopods, whose brain and behavior rival that of some vertebrates. The vertebrates, in turn, show a great variety of neural complexity in all groups, though in some—most notably the lungfish and salamanders—it has clearly regressed. Birds and mammals as a group have the highest brain-to-body mass ratios, though conspicuous enhancement is restricted to relatively few groups: the psittaciform and corvid birds, and the cetaceans, elephants, and primates among the mammals. Rather than viewing evolution of neural complexity and associated cognitive capacity as an inevitable and linear increase, a better interpretation of the facts is that nervous systems, the brains that control them, and the behavior that arises from them have evolved through natural selection to optimally adapt each organism to the niche it occupies.

Behavior is an energy consuming process. Some energy is expended in order to acquire more energy than can be extracted from the environment by remaining stationary. The short-term use of this energy is to promote the survival of the species within the strictures of its particular niche and lifestyle. Birds and mammals need a lot of energy just to maintain high body temperatures, enabling them to remain active at lower ambient temperatures for seeking mates and food, or keeping alert to danger. Predators need energy for chasing and capturing prey, as the latter need it for escaping. Animals build nests, dam rivers, dig burrows, and transport themselves by flying, swimming, walking, running, and hopping. Humans build houses and machines and toys, for shelter, work, and the social ends of play. All of these energy-consuming activities are geared toward the proximate aim of survival.

References and Notes

[1] Bullock, 1977

[2] Jerison, 1973

[3] Striedter, 2005

[4] Striedter, 2005.

[5] Butler & Hodos, 1996

[6] Jerison, 1973

[7] Northcutt, 1985; Striedter, 2005

[8] Jerison, 1973

[9] Butler & Hodos, 1996

[10] Striedter, 2005.

[11] Jerison, 1973

[12] Striedter, 2005.

[13] Striedter, 2005.

[14] Kappelman, 1996

[15] Striedter, 2005.

[16] Bullock, 1977

[17] Irwin & Schulze-Makuch, 2018

[18] Hodos, 1986

[19] Chaisson, 2013

[20] Friston, 2010; Neubauer and Fink, 2009

[21] Fornito et al., 2011; Fukusako, 2001; Schipul, Keller and Just, 2011

[22] Croxson et al., 2009; Kurniawan et al., 2010

[23] Day et al., 2010

[24] Rangel and Hare, 2010

[25] Buckholtz et al., 2010; Nestor et al., 2011; Knutson et al., 2001; Koob, 1992; Schultz, 2009

[26] Erlanson-Albertsson, 2005; Nestler, 2005; Nestor et al., 2011

[27] Ong and Mulhauser, 2011

6

Evolution and Information

To this point, we have developed the argument that the complexity of the universe has increased over time by forming local systems that resolve energy gradients more effectively. The total amount of free energy available within these local systems *and* their surroundings has decreased and entropy has increased, as required by the Second Law of Thermodynamics (SLT). Congruently, the local systems themselves have shown a net increase in complexity because they generally dissipate energy more efficiently and thoroughly than simple systems, and this is favored by the Principal of Least Action (PLA). Consequently, complexity and energy dissipation have thus shown a net tendency to increase together.[1]

As argued in Chapter 4, the rise of complexity can be quantified by an increase in the amount of information required to describe it. With the appearance of living organisms, local complexity achieved new heights, hence the role of information in channeling evolutionary trajectories took a quantum leap forward. Order, energy, and information became inextricably related.[2] The remainder of this book will deal with how, largely under human influence, the evolutionary process on our planet has undergone the transition from energy processing to *information* processing as its ultimate endpoint.

6.1 Information, Energy and Complexity

From the earliest femtoseconds of the universe, when pure energy first began to aggregate into matter, information became quantifiable as a descriptor of local systems, including their attributes, components, dynamics, and interactions. A brief review of how information pertains in the natural world will serve as a reminder of its pervasiveness since the beginning of time, and a prelude to our consideration of how it has become an increasingly consequential endpoint of evolution.

In the abiotic realm, systems involving optimization of physical flow, such as river channels, lava flows, ocean currents, and weather changes have been described as lawfully reflecting the optimization of energy transformations consistent with chance variations in the environment.[3] In these cases the physical parameters of the flows—volume, viscosity, velocity, gravitational gradients, etc.—provide an informational measure of their complexity. In the chemical realm, once atoms coalesced out of the chaos of the Big Bang, the universe consisted overwhelmingly of only two atomic species: hydrogen and helium, with traces of lithium and possibly boron. Successive generations of star formation have resulted in the evolution of more than 100 new chemical elements. Optimization of chemical combinations among this larger repertoire of atoms over eons has led to formation of more than a thousand new minerals.[4] This increase in chemical complexity has been accompanied by an increase in the amount of information inherent in the physical world. The evolution of these minerals occurred because chemical interactions were driven by energy gradients, which in turn released additional forms of energy for subsequent exploitation by new elements and minerals. These recombined under chance variations in temperature, pressure, chemical concentrations and the presence of the new elements and minerals. Each step and variable in this process of chemical evolution has added to the informational content of an increasingly complex physical world.

The same can be said for evolutionary changes from simple organic compounds into increasingly complex biochemicals in the prebiotic era, as described in Chapter 4. Not only did amino acids, carboxylic acids, lipids, and nucleotides constitute more complex molecules than their simpler precursors, but the metabolic pathways required to synthesize them grew in length and complexity. Each new step, each new enzyme required to catalyze a step transition, and every intermediate compound in the chain increased the complexity of the overall system, and hence the information required to, first, describe it, then after the innovation of genetic information, transfer and direct it.

Early in the history of life, certain molecular configurations became fixed in preference over others because those structures optimally facilitated the uptake and subsequent dissipation of free energy.[5]

For example carbohydrates have right-rotated configurations (Fig. 6.1), while those of polypeptides are left-rotated (Fig. 6.2). The non-randomness of this phenomenon, called homochirality, is a form of information conserved at the molecular level since the dawn of living cells. Over eons of protein evolution, specific foldable protein conformations became selected as energetically optimal for various critical functions. These functions included binding to appropriate ligands, catalyzing reactions, undergoing allosteric changes and avoiding aggregation. An information processing model of these functions has been used to successfully predict the optimal temperature range for the evolution of select families of proteins.[6] Congruently, five stages have been identified as likely steps in the evolution of the optimal structure of early protoproteins.[7] This dynamic process has been referred to as involving thermosynthesis,[8] primordial anabolism,[9] and autocatalysis.[10]

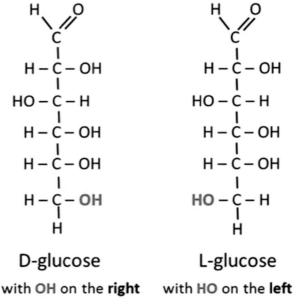

D-glucose
with OH on the **right**

L-glucose
with HO on the **left**

Fig. 6.1. Optical isomers of glucose. While both forms are found in nature, only the D-isomers of carbohydrates are metabolically active.

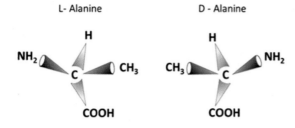

L- Alanine   D - Alanine

Fig. 6.2. Common amino acids are stereoisomers, or mirror images around a chiral carbon center. Amino acids can exist in either the D or L configuration. However, all chiral amino acids in proteins occur in the L-configuration.

At the other end of the size spectrum, the formation of stars and condensation of interstellar dust into planets and moons around them, and the aggregation of stars and their planetary systems into galaxies, represent a growth in complexity and information on a cosmic scale. Across the eons in all abiotic domains the same general conclusion has held; namely that increases in the rate density of energy transformation has resulted from a greater degradation of free energy per units of mass and time.[11] The amplification of energy rate densities by the evolution of life can clearly be seen in the increase in the energy rate densities in Figures 4.9 and 6.3.

Life forms exploit information processing to optimally dissipate energy. They do this by encoding the information necessary for all living processes in genes, including the ability to self-replicate (Chapter 4). Variations in the genetic store of information, by random mutations and other events related to the replication process, ensure that successive progeny have chance differences from each other in their informational instructions. This provides a quasi-random assortment of characteristics (phenotypes) for natural selection to operate on. Over the course of evolution, selection favors the survival and reproduction of individuals whose information allows them to transform energy more efficiently and completely. This has made it possible for life forms to potentiate the optimization process more rapidly than inanimate forms. Capitalizing on the acceleration of energy dissipation through natural selection of the information required to bring that

energy dissipation about, was the signature of the organic revolution.

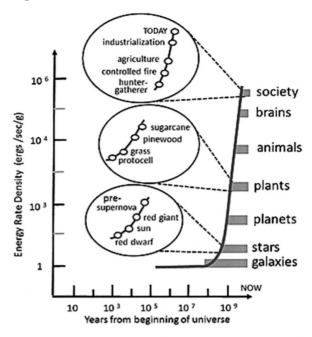

Fig. 6.3. Energy rate densities have increased dramatically in living compared to non-living entities. *Redrawn from Figures 30, 31, and 32 in Chaisson (2001).*

Successive life forms have become more efficient and effective at dissipating untapped sources of free energy, and in the process making new sources of free energy available. The survival advantages of being mobile, whether to approach food or avoid predation, was important for the initial appearance and later evolution of the nervous system (Chapter 5). The primitive nervous system facilitated information intake, processing, and distribution to remote body parts. The subsequent evolution of the nervous system continued that trend. Species evolved mechanisms for transducing information from the environment and internal processing of that information. These processes were optimal for exploiting the energy resources in their respective niches in keeping with the interactive effects of the PLA and SLT.

Significantly, the Principle of Least Action in physics and biology has its equivalent in the long recognized Principle of Least Effort in behavior.[12] The latter is ubiquitous in the coping behavior (instinctive or learned) of all types of organisms.[13] Humans brought their particular adaptations together to create a unique cognitive niche capable of exploiting energy resources to an unsurpassed degree.[14] Enhanced information acquisition and transfer was a key element in human success. Information processing was foundational to each of the three traits most responsible for human success: reasoning about causal effects, non-kin sociality, and communication.[15]

The remainder of the chapter will examine the psychological, social, and communicative ends of information consumption and utilization across species.

6.2 The Behavioral Uses of Information

Behavior refers to the actions of an organism and its responses to an environmental event or stimulus. Those behaviors acquired from experience during the individual's lifetime are learned. Conversely, behaviors that are instinctive are not learned, and thus are reliant on the genetic system for coding and transmission across generations.

One of the hallmarks of learning in animals is the ability to modify a prepotent response as a result of a brief learning experience. If the behavioral change is to stop performing an unlearned response after repeated stimulation, the learning experience is referred to as habituation. And if it is to increase the magnitude of the response, it is called sensitization. These are viewed as the simplest forms of learning. Habituation has been observed in all species studied.[16] Clearly, habituation is a form of information processing and its ubiquity across species argues that it has survival value across the animal kingdom.

Behavior akin to habituation can be observed in certain plants as well. For instance, a tropical plant, the mimosa, has an unlearned tendency to close its leaves when touched or shaken. Leaf folding is an adaptation to reduce the likelihood of animal herbivory. After a few seconds of repeated innocuous contacts the mimosa comes to ignore the contacts and stops curling its leaves, demonstrating habituation.[17]

Conditioned behavior in animals represents a more complex form of learning. In classical conditioning, unlearned (unconditioned) stimuli evoke unlearned (unconditioned) responses, but the organism can learn to substitute another stimulus (conditioned stimulus) to evoke the modified response (a conditioned response). For instrumental conditioning, any of a wide variety of stimuli can be learned to become a trigger for the organism to perform a behavioral act that is reinforced by being attractive (hence promoted) or aversive (hence avoided). The unlearned component is the attractiveness or aversiveness of the reinforcer. The learned component lies in the fact that a previously neutral stimulus is turned into a learned or secondary reinforcer.

Many behaviors are a combination of learned and unlearned actions. Whether learned or unlearned, behaviors are the product of the processing of information encoded within the organism or acquired from the environment and processed in a species-appropriate manner. Behaviors are phenotypes upon which natural selection acts. It is apparent that the survival of individual organisms is greatly enhanced by being able to learn to adapt to a changing environment. This requires importation and processing of new information. Learning thus complements evolution and provides a fine tuning of the evolutionary response to the challenges of nature.

Metacognition is the ability to be aware of and make judgments about one's own knowledge.[18] Examples include knowing whether something has been learned, judging how well it has been learned, or deciding if there is more to be learned. It can be inferred from the animal's increased time and effort to improve less certain knowledge. This information seeking behavior is similar to that for foraging for food. A range of species have been observed to exhibit metacognition, from bees,[19] to rats,[20] to humans.[21]

Information seeking behavior has been shown to increase as the complexity of the reward site increased, or as the amount of information to be gained increased, and decreased as the information to be gained decreased. For example rats were observed to work to reduce the uncertainty about the location of food. The information did not alter the likelihood that the rats would receive the food but did allow them to obtain the food with less effort in a shorter time.[22] Critically this observation exemplifies the close relationship between foraging for information and foraging for energy (food) under the PLA's demand for optimality.

The example in bees is instructive because it occurred in a non-laboratory setting without training. When leaving a new food source, bees spent increased time and effort examining any food site which was a new or a more complex site before leaving, and less time orienting before leaving a familiar site.[23] This shows that uncertainty about information was enough to influence behavior.

Across species, humans stand out as the most cognitively equipped to engage in information foraging. Prolonged childhood and adolescent learning clearly allow a period of intense training and learning. Consequently it seems reasonable to speculate that foraging for information in humans may have evolved from foraging for food, and thus may exhibit some of the same features.[24] Human behavior in searching through "patches" of information shares a number of characteristics with animal food foraging, including the use of cues to optimize time allocation, using a win-shift strategy to avoid already visited patches, using a just-right strategy to forage easiest to access patches first, using a stop rule to decide when diminishing returns from a patch indicate shifting to another patch, using a tradeoff rule when juggling priorities that compete with foraging, such as using a value rule to guide the choice of which patch is the most "nutritious," and finally exhibiting individual differences in foraging expertise. In addition, species vary in the degree to which they use metacognition to optimize foraging. Given the conservative bias in evolution to retrofit prior adaptations to new challenges, it seems reasonable to conclude that the cognitive underpinnings of human information processing may have been adopted from a suite of cognitive biases which supported human food foraging.

Another form of metacognition which treats information as energy is causal intervention. The informational uncertainty in this case is causation. Crows can acquire new information produced by their own activity. For example, through repeated testing of a tool to capture prey, they can refine the tool into a more effective instrument (see Section 7.1). Yet, they are unable to spontaneously test a

sequence of events for causality. In humans, the ability to do this readily may be related to an inherent cognitive bias to perceive causal relations between events even when there are none. This human bias in cognition regarding causation is seen across cultures and includes the tendency to focus on the possibility of human agency causing the event.[25] Given that humans are highly social, a bias to perceive unexplained events as possibly caused by a human agent is not surprising. Of course it can also be a cognitive trap when it leads to belief in unfounded conspiracies or the presumption of agency by imagined causes, such as supernatural forces. Combined with the inordinate human social curiosity about the perception of others, these variations in metaphorical perspectives have led to the reasonable conjecture that they provide the well-developed human cognitive mechanisms for understanding the perspective of others (Theory of Mind), assessing probable causal sequences, effective planning, and abstract thinking.[26]

6.3 The Social and Communicative Uses of Information

Multicellularity can be viewed as the simplest form of social interaction. In general it evolved in cases in which it increased energy dissipation beyond dissipation by the single-celled form. Yeast clearly reflect this principal by switching from a unicellular to a multicellular form when enhanced energy gradients permit.[27] In more complex organisms the evolutionary advent of multicellularity improved overall functioning through integration of specialized cells. The integration and attendant need for intercellular communication led to species with increasing numbers of genes and gene interactions. This can be seen in the dramatic increase in the number of protein coding genes and regulatory genes (Chapter 4, Figure 4.8) over the course of evolution of plants from algae-like predecessors. This increased flow of information is paralleled by a 20-fold increase in energy transduction across plant taxa.[28]

An auxin transport system in plants mediates plant growth vectors sensitive to light and gravity. A second intercellular communication system in plants allows them to cope with pathogens and parasite attacks in a manner equivalent to the coordinated response of the immune system in animals. In plants part of this response is also to form mutualistic protective relations with other organisms, including bacteria, fungi, and insects.[29] From the foregoing account it is apparent that plants, even though they lack nervous systems, have developed functional forms of "social" coping that critically involve information processing.

Some plants are capable of communicating warnings of imminent drought threat to other plants, enabling them to take pre-emptive anticipatory action.[30] The plants forewarned of drought stress were able to adopt measures to reduce drought susceptibility. More importantly, the spared plants subsequently shared their soil rooting volume with their drought-stressed neighbors, assisting in their post drought recovery. Plants can also inform each other about an imminent herbivore attack by liberating chemicals which signal to adjacent plants that they should begin producing defensive chemicals. These warning chemicals are reacted to more strongly by plants closely related to the attacked signaling plant than by more distantly related members of the same species.[31] Similarly plants can use the signals from neighbors to avoid competing with closely related plants for resources or to anticipate competition from unrelated species.[32] Thus roots are not only able to respond differentially to self and non-self, but also to distinguish between self and other plants regarding their degree of relatedness.

Beyond feeding nutritious nectar to pollinating animals (insects, birds, bats), and feeding fruit to seed-distributing animals, plants have acquired complex forms of bilateral cooperation with insects. For example, some species of acacia have evolved a form of bilaterally advantageous mutualism with ants. The ants take up residence in the acacia and consume its nectar. [33] When a herbivore begins eating the acacia, the stinging ants stop it. Another example of mutualism is the tobacco plant's reliance on a species of wasp for protection against budworm caterpillars. When attacked by the caterpillar, the plant releases a chemical which attracts a parasitic wasp. The wasp lays eggs inside the caterpillar, which hatch into larvae that consume the caterpillar.

The chemical attractant can be tailored to attract wasps which prey on a specific species of caterpillar.[34] Plants are also able to release volatiles which repel females of foraging caterpillar species.[35]

We have already mentioned the importance of interspecies mutualisms in animals relating to their interactions with insects and plants. From this we would expect that animals would also have mutualisms with other animals of both kin and non-kin varieties. This is indeed the case. A good example of the latter is the cleaner-host relationship among fish species. Wrasse cleaner fish clean the skins of many species of host reef-fish. The relationship is not a static one. Some cleaner fish learn to cheat the host by using the host for protection but offer little cleaning in return.[36] Long-nosed parrot fish change cleaners who cheat if there are other cleaners available. However hosts with only one cleaning partner available, punish their cheating partners.[37] Two points can be made. Clearly there is information processing here represented in the genetically inherited cooperative relation between cleaners and host fish. In addition the learning component is apparent in the complex conditional variations in the behavior of the cleaners and hosts.

The same circumstance-specific sensitivity is seen in non-kin mutualism studies of humans. Both of the following studies illustrate the importance of the availability of a satisfactory alternative in expressing the conditional nature of dependencies.

Students in social network role playing games were less likely to change cheating partners if the costs of doing so in time and effort were high.[38] If the option to stop collaborating was clearly available yet expensive, this in itself was likely to increase the level of cooperative behavior. The mere presence of an option to end a partnership was enough to encourage cooperation.

The second study generated mathematical models to illuminate the transition of Neolithic hunter-gatherers to proto-herding and farming.[39] This was an interesting change because the current anthropological view is that activities of the herders were quite loosely organized and egalitarian whereas the farmers were hierarchical and tended to despotism. It is true that nomadic hunter-gatherers often joined in large collaborative efforts. For example, many hunts required prolonged group effort—the building and maintenance of traps, coordination of herding, slaughtering, and storing of meat and hide preservation of large prey such as mammoths, buffalo, caribou, and whales. However the suzerainty of the group's chief was restricted by tribal dispersal during the off season. There were a few cases in which new technology permitted year-round access to a major resource. In such cases, despotism and accompanying sequestration of resources created a wealthy hierarchy and an authoritarian regime, such as the development of seafaring boats for hunting whales.[40] So the question remains, if a regime became despotic with punitive controls and unequal distribution of wealth, why would an individual voluntarily trade an egalitarian regime for this? The modelling clearly showed that the transition from Neolithic egalitarian to despotic hierarchical groups began when leader-facilitated group effort led to a significant increase in resources and a subsequent increase in group population size. This gave the survival advantage to members in the larger more productive group. However defections still occurred. But when the growth in productivity and population was accompanied by reduced availability or increased cost of alternative resources, this made alternative life styles less attractive and defections dropped off.[41]

A similar interplay of factors is clearly involved in all of these examples of decision making, which ultimately lead to greater energy dissipation even though the proximal mechanisms are obviously different. As energy consumption in ancestral human populations increased, social organizations became more complex, technology was invented, and cultural differences between isolated groups increased. The complicated interplay of energy, information, and technology as human societies grew more complex and generated cultural attributes will be explored in our next chapter.

Summary of Chapter 6

The optimization of energy dissipation, as mandated by the PLA and constrained by the SLT, has resulted in the aggregation of increasingly complex local systems over time, with a corresponding rise in the informational content and processing within those systems. The topographical complexity that arises from flowing systems, such as river channels and lava flows as they cut new paths across a planetary surface in the process of dissipating potential energy gradients, is an example of the interplay of energy and information. The same exploitation of optimal energy and informational gradients can be found in the evolution of chemical elements and molecules, the appearance of chemical catalysts, the appearance of homochirality in specific types of molecules, and the formation of optimally folded proteins. The evolution of larger, more complex, multicellular forms of life has optimized information processing and energy dissipation in a radically different way— by enabling the acquisition of information during the life time of individuals (selective reinforcement), and by encoding information of survival value into a genetic repository that maintains adaptive innovations across generations (selective survival).

Sensitization and habituation are the simplest forms of acquired behavior, and are seen in most species of animals and some plants. Many animal behaviors are a combination of learned and inherited components, as seen in classical and instrumental conditioning, animal predispositions in learning, human cognitive biases, and epigenetically inherited tendencies. Metacognition—the ability to make judgments based on one's own knowledge—is one of the more highly evolved forms of learning, and one that provides a clear example of the value of optimizing information processing. Testimony to the value of metacognitions is seen in the wide range of species that make use of them, from bees to humans.

Though lacking a nervous system, plants still have the capacity to emit and detect information through chemical signals and other means. That they do so, and thereby promote the reproductive success of their species and enhance their ability to consume more energy, is shown by a variety of phenomena, including conspecific warnings about encroaching herbivores, response to the threat of drought, territoriality, and competition for resources. Sensitivity to genetic relatedness even allows plants to exhibit cooperative kin selection and out-group biases characteristic of animals. As noted in Chapter 4, the value of increased information processing in plants over evolutionary time is exemplified by the increase in protein coding genes and a concomitant increase in energy transduction.

In animals, some species are more social than others, but all rely on information transmission between individuals to a degree. Cooperation among kin is commonly observed for protection, feeding and training of young, and cooperative hunting and colony protection. Animals also exhibit many forms of mutualism, which requires transmission of information across species boundaries. Adjustment to the degree of cooperation between species can be quite sophisticated, depending on the availability of alternative partners. Adjustment of mutualistic partnerships in fish have similar forms of modulation to those observed in humans. Both are sensitive to the relative cost/benefit of maintaining versus leaving the relationship.

One of the most obvious instances of animal behavior critical for energy transformation is foraging for food. Given the bias of the evolutionary process to make do with existing systems, it seems reasonable to speculate that foraging for information might have similar characteristics to foraging for food. This is indeed the case, as many features of foraging for food are duplicated in foraging for information. Thus in the evolution of our behavioral biases, we have embedded the interaction of energy and information in our genetic inheritance.

References and Notes

[1] Karnani, Pääkkönen & Annila, 2009

[2] Herrmann-Pillatha & Salthe, 2011

[3] Bejan & Lorente, 2011

[4] Hazen, 2010

[5] Jaakkola, Sharma & Annila, 2009

[6] Morcos et al., 2014

[7] Trifonova & Berezovsky, 2002

[8] Muller, 1996

[9] Kaufman, 2009

[10] Ingber, 2000

[11] Chaisson, 2010

[12] Mowrer & Jones, 1943; Zipf, 1949

[13] Breland & Breland, 1961; Bejan & Marlen, 2006

[14] Tetlock, 1983; Kaila and Annila, 2008; Selinger et al., 2015

[15] Pinker, 2010

[16] Rankin et al., 2009

[17] Gagliano et al., 2014. As expected, the reduced leaf curling behavior was acquired more quickly in a low-light environment because low-light puts a premium on the leaves' being open to capture as much light for as long as possible.

[18] Metcalfe, 2008

[19] Lehrer, 1993

[20] Kirk, McMillan & Roberts, 2014

[21] Call & Carpenter, 2001

[22] Kirk, McMillan & Roberts, 2014

[23] Lehrer, 1993

[24] Metcalfe & Jacobs, 2009

[25] Hewstone, 1989

[26] Gallese, 2007; Pinker, 2010; Hartung & Renner, 2013; Douglas & Muscovici, 2015

[27] Annila & Annila, 2008

[28] Chaisson, 2010

[29] Baluška, Volkmann & Menzel, 2005

[30] Falik et al., 2012

[31] Heil & Karban, 2010

[32] Dudley & File, 2007

[33] Goheen & Palmer, 2010

[34] De Moraes et al., 1998

[35] De Moraes, Mescher & Tumlinson, 2001

[36] Bshary & Schaffer, 2002

[37] Bshary & Grutter, 2002

[38] Bednarik, Fehl & Semmann, 2014

[39] Powers & Lehmann., 2014

[40] Arnold, 1995b

[41] Powers & Lehmann, 2014

7

Evolution of Society and Culture

The evolution of life on earth has been shaping ever more effective and efficient forms of energy acquisition and consumption for over three billion years, in response to the imperative of the Principle of Least Action (PLA), in the direction mandated by the Second Law of Thermodynamics (SLT). During the Cenozoic Era (most recent 65 million years), mammals and birds have been particularly high energy consumers by virtue of their warm internal body temperatures and generally high levels of activity. For close to six million years, a branch of primates has been evolving through a number of related species characterized by an increasingly upright posture and reduced dentition. This is the human (Hominin) line of Great Apes. For most of their history, hominins have evolved in synchrony with other Primates as primarily forest-dwellers consuming an omnivorous diet, relying on strong diurnal stereoscopic vision, and using their hands for grasping and fine manipulative control. Within the last two million years, ancestral humans, and now *Homo sapiens*, the only surviving species of humans,

have emerged as the supreme energy consumers and dissipaters in the living world. Most obviously this is expressed in our greatly enhanced and technologically driven consumption of fuel, and more recently, in our creation, voracious consumption, and exchange of information—the greatest singular emergent manifestation of energy transduction in the history of life on our planet. In this chapter we explore the uniquely human forms of energy consumption and its consequences, beginning with the evolution of material technology.

7.1 Origin of Material Technology

Technology can be defined as the use of energy, tools, material, and information to amplify the impact of a species on its environment.[1] The precursors of technology can thus be found in the use of tools, which leverage an animal's ability to interact with its environment.

7.1.1 Animal Antecedents to Tool Use

True tools are objects that an animal detaches from the substrate and manipulates to serve a purpose. Tools provide one of the more compelling examples of evolved increase in energy transformation. Of direct relevance to the energy dissipation hypothesis, most of these tools are involved in food extraction, storage, or consumption.

Examples of such tool use among mammals and birds abound. Otters use rocks to break open mollusks; chimpanzees and capuchin monkeys use rocks to crack nuts and shells; dolphins use marine sponges placed over their snouts to probe for fish on the sea floor. Elephants use their versatile trunks for fashioning tree branches into flyswatters and probes, hurling objects at other animals, and stacking objects to extend their vertical reach. Various primates dig up tubers with sticks, retrieve honey from hives, and use sticks to fish for termites. Chimpanzees use crushed leaves to sponge up food and fluids. Tool use has been documented in over 30 bird families—most prominently among the psittaciformes (parrots, lorikeets, and cockatoos) and corvids (crows, ravens, jays, and magpies).[2]

Why do animals use tools in foraging for food? After all, depending on the species and the type of food, proficiency in tool use requires time in searching for and shaping the tool, and learning to use it. Evidence suggests that animal tool use contributes directly to a gain in energy intake beyond that afforded by simply consuming staple foods. New Caledonian crows can get over 80% of their required daily protein and lipid intake from two foods: long-horn beetle larva and candlenut tree nuts, even though these foods form only a fraction of their daily intake.[3] Both the nuts and larvae are obtained with tools. The crows use sticks to tease the larvae, which bite onto the stick and are pulled from their burrow. The nuts are cracked by dropping them on a hard surface. Both forms of tool use require considerable practice for mastery by juveniles. The crows obtain other food—lizards, snails, carrion, and fruit, for example—without using tools. However, the use of tools is instrumental in efficiently harvesting the greatest amount of food energy available.

Not all tool use by animals is explicitly related to resource acquisition or consumption. Construction of homes, dens, nests, and shelter all involve manipulation of materials, and further represent tool use by non-human animals. Examples include insect (ant, bee, and wasp) nests, fish nests, bird nests and bowers, holes dug by invertebrates and vertebrates for dens, branches cut by beavers for lodges and dams, and the bent and broken branches used by chimpanzees in building nests. Elaborate nests constitute a component of courtship display in bower birds. Hermit crabs and octopi use transported shells to provide refuge from predators. Self-decoration with extraneous material in many species of vertebrates and invertebrates is thought to have similar antipredator survival value, except in humans where self-decoration functions in social interactions.[4] Many animals cache supplies and food for later use and consumption. This use of tools and materials is similar to the use of construction materials in that it protects a resource but is not explicitly involved in acquisition or consumption. Note, however, that all these uses of tools indirectly promote energy resource acquisition and consumption, in the sense that they allow the animal to survive for another day in security and with shelter while resting, hibernating, avoiding predators and caring for young.

7.1.2 Human Use of Tools

The use of tools by humans, then, is clearly an extension of a behavioral trait shared by many animals, most primates, and all the great apes. The opportunistic use of tools, by which implements are found and used with little modification where they are needed, would have been inherited from proto-human ancestors. The first artefacts indicating the human use of bones as tools date back to 3.4 Mya, flaked stone tools to 2.6 Mya, and bifacial stone tools to 1.7 Mya. Composite tools, such as stone points fastened to a wooden shaft (the process known as "hafting") appeared around a half million years ago. A critical element in the appearance of bifacial stone tools was the evolution of the human hand with modern grasping abilities. The critical bones for maintaining a power grip and a precision-pinching grasp have now been identified in hand fossil bones dated at 1.42 Mya, likely from *Homo erectus*.[5]

The earliest members of the genus *Homo* appeared during the mid- to late-Pliocene,[6] over 2 Mya (Table 7.1). As the climate in Africa grew drier and less forested, the diet of these ancestral humans began to incorporate more meat, probably scavenged from larger mammalian carcasses. To butcher the carcasses, more sophisticated cutting tools, reliably available, would have been needed. Not surprisingly, then, the oldest deliberately flaked stone tools have been found among skeletons of large animals, from sites dated about 2.6 Mya.

The deliberate creation of stone tools by humans marked the beginning of the Stone Age, which conventionally is divided into Paleolithic ('old'), Mesolithic ('middle'), and Neolithic ('new') periods.[7] By far, the longest period is the Paleolithic, during which stone tools gradually became more sophisticated, through a series of what archeologists call "industries."[8] The oldest and longest lasting of these was the Oldowan Industry, characterized by production of unifacial cutters and scrapers. The first human species to fashion Oldowan tools was probably *Homo habilis* in the late Pliocene (Table 7.1).

Table 7.1 - Temporal correlation of environmental, cultural, and technological changes related to the evolution of humans during the Neogene (Pliocene – present). Events are listed according to the approximate time of their first appearance.

Geology Epoch / Mya	Ecology	Anthropology	Archeology (Industry)	Technology
Holocene 0.01	Interglacial, warming	*Homo sapiens* - sole surviving hominin (1450 cc) **	*Neolithic (Magdalenean)*	Iron Age Copper/Bronze Age Clovis/Folsom points
Pleistocene 1.0 1.8	Wurm/Wisconsin* Riss/Illinoisian* Mindel/Pre-Illinoisian* Günz/Pre-Illinoisian* Pre-Pastonian	*H. neaderthalensis* (1500 cc) *H. sapiens/H. helmei* (1400 cc) *H. heidelbergensis* (1200 cc)	*Mesolithic* (Mousterian) Art, decoration *Upper Paleolithic* (Acheulian)	Bows and arrows Triangular flakes Hafting Initiation of FIRE Spear throwers Control of FIRE Bifacial hand axes and cleavers Spears
2.0 Pliocene 3.0 4.0 5.0	Cool Grasslands spread in Africa Panama Isthmus forms; ocean currents alter Forests decline; grasslands spread in Africa/Asia	*H. erectus* (975 cc) *Homo habilis* (700 cc) *P. robustus* *Paranthropus bosei* (500 cc) *A. africanus* (450 cc) *A. afarensis* (400 cc) "Lucy" *Ardipithecus ramidus*	(Oldowan) *Lower Paleolithic*	Unifacial cutters and scrapers

* Alpine/North American glaciations   ** brain volume

Homo habilis was succeeded by several lineages related to Homo erectus, some of which migrated out of their continent of origin in Africa into Eurasia in the early Pleistocene. At least four glacial advances and retreats beginning in the mid-Pleistocene presented adaptive challenges to these descendants from neotropical ancestors. Each warming during interglacial periods provided opportunities for range expansions and the development of more efficient and effective tools and weapons. Bifacial cutting tools of the Acheulian Industry enabled the sharpening of heavier sticks into spears. The effectiveness of spears was enhanced by spear-throwers, and by hafting sharp, bifacial stone points to wood. The primitive Oldowan hand cutters evolved into bifacial Acheulian cleavers, which created a formidable chopping tool and weapon when hafted onto a wooden handle. By the late Pleistocene, bows and arrows had been invented, and smaller, precise arrow points were created in the Mousterian Industry (Table 7.1)

7.1.3 Origin of Human Habitations

Human construction of habitations to provide a controlled internal environment is an ancient activity extending back to late Paleolithic times.[9] Hunter-gatherers of the late Pleistocene moved about and therefore had no use for permanent structures. Natural products, such as leaves, limbs, and animal skins would have been the earliest materials for constructing habitations. More durable natural products, like clay, stone, and timber came into use with the advent of agriculture, which enabled more stationary living. Circular stone formations believed to have been the base of some type of housing have been dated back to more than 12,000 years ago. One of the more interesting building materials dating back to antiquity is bitumen. Noted for its waterproofing, adhesive and malleable properties, it was used (along with beeswax and conifer resin) in building mortar, jars, cisterns, pipes, boat hulls, tool handles, sculptures, stamp seals, and as a balm in mummification. One of the earliest uses discovered to date is in the hafting of stone heads to axe handles by *Homo neaderthalensis*.[10]

7.2 Control of Fire

The advent of the control of fire by early hominins signaled a paradigmatic shift in human evolution—a singularity in the evolution of effectiveness with which energy could be transformed and consumed. The habitual use of fire dates back 300,000-400,000 years,[11] but evidence of concentrated fire use in the form of scorched earth associated with carbonized seeds and nuts is found much earlier (780,000 years).[12] No other animals have been observed to control or use fire, so the mastery of fire is a singular accomplishment of the human species.

The use of fire by humans extended home ranges into otherwise inhospitable environments, afforded warmth and light, offered protection against nocturnal predation, and ultimately aided in the construction and use of more and better tools. While provision of warmth and protection from wild animals and possibly other primates are obvious advantageous uses of fire, the most far-reaching consequence of controlled fire was for cooking food.[13] Cooking made many more foods digestible, reduced the energy costs of digestion, decreased the time for digestion, and increased the caloric yield of food.

Cooked food was preferred partly because it increased the availability of preferred substances that tasted of sweetness, tannin, and umami (glutamate taste–signaled protein).[14] Cooking also significantly raised the caloric yield of food (from 30 to 200% depending on the food), reduced food toughness, enhanced food safety, increased the varieties of food that could be consumed, and reduced the energy cost of digestion (by 30%) and time required for food digestion (by 300%).[15] The biological consequences of cooking included the evolution of a smaller gut by 40%, reduced jaw and size of teeth, and a larger more energy demanding brain.[16] Brain energy consumption increased 3-fold from the 300 cc brain of early *Australopithecines* to the 1500 cc brain of *Homo neaderthalensis*. Bipedal gait which initially preceded brain enlargement was further facilitated by a smaller gut, and likely also by the advantages of hands freed for carrying, controlling fire, and gestural communication, as well as readier access to underwater plant food sources.[17] These adaptations in turn enabled larger more complex social networks which in turn fed back to further increase brain size.[18] No human societies live without cooking,[19] and other species prefer cooked over uncooked food.[20]

In addition to providing warmth, light and defense against predation, the control of fire was in time applied to more sophisticated non-cooking activities, such as clearing land, hardening of pottery and weapons, stone flaking, glass making, masonry, and — following the Stone Age — metal refining and energy for the manufacturing processes of the industrial and postindustrial eras.[21]

7.3 Origin of Human Language

Language is the critical medium of communication in humans, so the evolution of language is a seminal event in the history of the human species and the fate of the entire biosphere on Earth.

We frankly don't know enough about communication in other animals to judge the extent to which they communicate through proto-languages, though Neubauer's 2013 review of communication in four species with large brain-to-body ratios suggests that information transmission may be more extensive than we realize.[22] We will take the direct approach of comparing communication in humans with our closest living relatives among the primates, especially the chimpanzee.

Human language is quite different from oral communication in nonhuman primates. Vocalizations in monkeys and apes have limited variation, are mostly fixed at birth, show minimal change in development, and little later individual variation. Most animal vocalizations are typically involuntary reflections of current emotional states,[23] compared to the large number of variations of facial expressions, hand gestures and bodily postures that apes use for communication.[24] In contrast with vocalizations, ape hand gestures are under volitional control, being largely learned and so qualifying as a form of intentional communication. Their meaning can change depending on the context, and they take into account the attentional state of the receiver. Overall, the comparative primate literature supports the view that human oral communication evolved from the primate system of controlling gestures.[25]

Speech is superior to gesture, however, in being effective in the dark, in not requiring directed visual gaze of either the sender or recipient, and in freeing the hands of the sender for other tasks.[26]

Additional evidence that speech evolved from gestures comes from neuroscience and archeology. Regarding hemispheric laterality, both human gestural and chimp gestural communication show right hand preference which indicates control by the left hemisphere in most individuals.[27] Detailed evolutionary modelling has shown a shift in the neuroanatomical changes leading toward lateralized motor functioning coincident with the ape ancestral split from other primates (approximately 10 Mya), with accentuation in the human ancestral lineage (approximately 6 Mya).[28] Asymmetric cerebral control of human gestural bias is evinced in signing by the deaf, manual movements when people are talking, dance movements, and pointing gestures by infants.[29] The hemispheric bias is also true of speech in humans but to a slightly lesser degree. Larger left hemisphere planum temporale structures are associated with individual right-handed gestural preferences in humans and chimpanzees.[30] Speech and gestures involve equivalent cortical areas. For example, the F5 area (identified by characteristic nerve cell types) of the premotor cortex of monkeys contains mirror neurons that allow monkeys to recognize the actions of others and control their own gestural movements. The monkeys' F5 area is homologous to Broca's area in humans (Brodmann area 44) which controls speech and complex manual movements.[31]

The hypothesis that grasping objects is evolutionarily linked to grasping meaning through shared neuroanatomical structures is supported by paleontological evidence based on changes in fossil skulls suggesting that stages in the evolution of language—the simple imitative largely gestural period, the more complex early oral period, and the final emergence of vocal language—paralleled the evolution of increasing sophistication of human tool making.[32]

Early tools were primarily used for procuring food. With the human control of fire, and later the ability to initiate it, the human diet shifted to a greater reliance on meat and the scavenging or hunting of larger animals. Larger animals yielded more nutritional value and energy per unit of effort. Not only did these activities require greater coordination and cooperation among individuals, but the increasing sophistication of tools and the socialization associated with cooking and life built around the use of fire, all demanded the increased sophistication and precision of information processing that language provided. The evolution of language, spurred by the need for food and the uses of fire,[33] thus ties energy consumption directly to information processing in our human ancestors. Taken together, these effects were manifest in the development of language and a familial division of labor, with females tending the hearth, gathering and preparing food, and caring for offspring, and males defending the hearth and hunting game. In time, a more elaborate division of labor within the camp group could be implemented for the manufacture of shelter, clothing, utensils and weapons, and herbal preparations.[34]

In all probability, tool use, control of fire, cooking and increased caloric consumption, brain enlargement, co-ordinated social activity, and language interacted and co-evolved in a synergistic way, as humans dealt with the challenges of oscillating climate cycles both in Africa and Eurasia. Taken together, these factors constitute what has been referred to as the "human adaptive complex" (Fig. 7.1),[35] which underlies much of the thinking about why the human brain underwent such an explosive expansion, beginning about two million years ago.

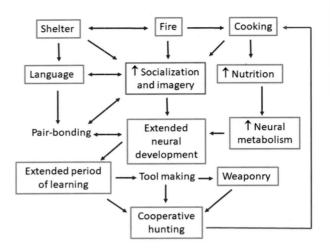

Fig. 7.1. The "Human Adaptive Complex." Ecological, social, cognitive, and historical factors have interacted in a complex way to influence the evolution of human brain and behavior, with information (through language and learning) playing a growing role. *Adapted from Kaplan et al., 2007)*

Taken together, elements of the human adaptive complex have vastly increased energy utilization; and the evolution of a brain which maximizes its consumption increases human fitness, and therefore is favored by natural selection, thereby setting up a positive feedback loop for enhancing the unique qualities and capabilities of the human brain (Fig. 7.2).

Energy Utilization Feedback Loop

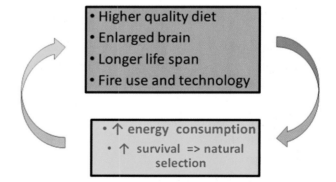

Fig. 7.2. Energy utilization feedback loop. Elements of the Human Adaptive Complex promote greater energy utilization, which has adaptive advantages for survival, reinforcing those adaptations.

7.4. Art and Culture

Culture is fundamentally an informational system that allows individuals to take advantage of collective action in securing their energetic needs.[36] It provides social cohesion, reinforces collective values, and promotes collaboration toward long-term goals. And of particular relevance to our central thesis, it results in more effective energy use by individuals within a cooperative group than by individuals operating alone.

As noted previously, the evolution of humans as the most successful species in transforming energy depended in large part on the social development of the human brain, the emergence of language, and the ability to create and use tools—all combining to lead eventually to the development of technology. Culture may be viewed as an informational system making it possible for individuals to take advantage of the exchange of information. Cultures express their collective values, beliefs, and norms through material creations such as art and crafts, through their oral and, ultimately, written language, and through behaviors such as music, dance, games, and rituals.

Language is widely regarded as a key element in the development of human culture.[37] The evolution of human language from gesture and inarticulate vocalizations provided a basis for the accumulation, storage and transmission of information in oral and eventually in written form.[38] Language permitted the conscious review, reconstruction, and planning of an individual's experiences, and the sharing of those experiences with others.[39] Complex, effective group action was thereby greatly facilitated.

The development of symbolic language also enabled the transmission of more extended narratives of current and past events. Narrative stories facilitate construction, recall, and relaying of a complex series of events which may be considered for their novelty, cultural specificity, observational acuity, logical sense and causal plausibility. Beyond the theatrical and ceremonial occasions, the story form is used in narrative lessons of instruction, social gossip, and childhood role playing games. Stories bind together members of the group in a tradition across individuals and time. The benefits of group membership include anxiety reduction, acquiring new information, enhanced social bonding and above all increased productivity.[40]

Over the course of human history, art form became encapsulated in universally ritualized communal celebrations involving visual decoration, storytelling, music, dance, drama and feasting. Audience participation in the artist's performance is often expected. Group drumming, dancing, singing and the theatrical performance of narratives are obvious examples with likely subsequent increased social bonding and motivation for collective action, such as warfare, building projects, and farming. The commitment by group members to group projects requiring extreme effort and even self-sacrifice are ensured by the engagement of members in group rituals involving intense emotional experiences.[41]

Fig. 7.3. From the wall of a dimly lit cave formed by forces tracing to the start of the universe, a product of the evolutionary imperative leaves fragments of information for the ages to come. *Art by Charles Beck*

Within the context of our central thesis, culture can be viewed as an elaborated and extended form of social organization and behavior that consumes more energy more effectively than otherwise would occur. In other words, energy dissipation by individuals as group members is greater than that of individuals operating alone.[42] This can be seen at several levels.

The concept of the transformation of energy has been formally applied to linguistic communication.[43] The opportunities for group energy expenditure were greatly enhanced by the ability of members to describe complex conditional contingencies between imagined events.[44] For ancestral cultures this included the prediction of celestial and seasonal events as well as preparation for planting, harvesting, storing and distribution of grain. For modern human societies, this extends to every facet of collaborative human endeavor, from team sports to complex construction of infrastructure, to mechanized warfare.

Art forms also serve to increase the rate of energy transfer in human groups compared to the group member working individually. Evidence for the role of art in energy transfer can be seen in the generation of new understanding and raised emotions within the audience.[45] The highest form of personal gratification has been described as being engrossed in some activity, immersed in a state of flow—a feeling of radiating energy, losing oneself in an activity, during which awareness is suspended, and the flow of time is altered.[46] A neural measure of appreciation is increased activation of the medial prefrontal cortex and dopaminergic reward circuits of the brain while the respondent anticipates and receives reward during exposure to a work of art or performance.[47] Activation of the dopaminergic circuits results in increased utilization of brain glucose.[48] To sum up, salient features of human culture including language, artistic expression and appreciation, spiritual experiences, and life satisfaction may be seen as manifestations of effective energy transformation, even if this plays no conscious role in the motivation of these activities.

Manifestations of energy transfer can take many forms, whether in the brain, in the individual's activity, or in concerted group action, such as building monuments, fighting battles, growing crops, and celebrating harvests. At a personal spiritual level, communal celebratory events are often accompanied by subjective sensations of a transcending flow of energy, of surrendering oneself to a greater whole, and of experiencing a spiritual dimension of being at one with the universe, expressed in the belief that we belong to a greater whole and that life is common in the cosmos.

In addition to the additive capacity for energy consumption attendant to cultural behavior and activities, it should be noted that changes in the behavior of cultures can occur much more rapidly through adoption and learning of new norms, and can be more finely tuned to environmental pressures than changes in behavior resulting from natural selection acting on genetically mediated behaviors. In turn this enables individuals to join in collective action to secure group energetic needs. Thus, effective use of energy is not only enhanced by culture, but can be adjusted much more quickly in response to demand within a cultural context.

7.5. Agriculture and Animal Domestication

Domestication involves changes in the genetics (usually through selective breeding) of a species, either plant or animal, to enhance traits beneficial to humans. Hunter-gatherers are human populations living primarily on undomesticated species of plants and animals. *Homo sapiens* have been hunter-gatherers to some degree for 90% of their history. The beginnings of domestication of animals may go back 100,000 years, with the certainty that wolf/dog domestication was underway at least 30,000 years ago.[49] The earliest forms of wolf domestication probably involved self-selection by individual wolves tame enough to tolerate proximity to humans while foraging human-slaughtered carcasses. While other large carnivores have been individually tamed, none have been successfully domesticated. Other animal species undergoing early domestication (prior to 8,000 years ago) included sheep, pigs, goats and cattle.[50]

There is evidence that major cultural shifts occurred in modern humans as a result of early animal domestication. Specifically there is evidence that modern humans evolved from being adventitious scavengers of vulnerable prey to being mobile long-distance hunters of large healthy prey in the mid- to late Paleolithic. The initial domestication of canines in the late Paleolithic (35,000 years ago) was concurrent with the appearance of mammoth skeleton mega sites.[51] Genetic analyses confirm that the canines at those sites differ genetically from ancient wolves, modern wolves, and dogs.[52] The large number of bones at these sites (some in excess of 100 mammoths) indicate a significant increase in the number of kills and a consequent higher meat yield than at sites occupied by humans without dogs. Faster population growth in canine-using groups of hunters would be expected due to improved nutrition and reduced energy costs. The site size increased as did the number of stone tools and the meat yield estimated from the number of prey. The territorial instincts of canids make it probable that the hunters retained control of the large kill sites more effectively as a result of the instinctive guarding by dogs of carcasses from marauding scavengers. This conjecture is corroborated by the larger number of undomesticated canine remains found at these sites than heretofore anticipated. Some of these mammoth megasites had dwellings framed by mammoth bones and featured hearths indicating the longer duration of tenancy at these sites. While contemporary Neanderthal hunters employed entrapment and ambushes in capturing large prey, the *Homo sapiens* of the upper Paleolithic introduced portable projectile weapons capable of killing at a distance.[53] Combined with the hunting alliance with canines, these innovations in weaponry afforded the hunters advantages in economy, safety and mobility over the Neanderthals, resulting in the observed large increase in the size and number of mammoth kill sites.

The implications of canine domestication and the development of projectile hunting for the SLT and PLA are apparent in the consequent increased availability of meat as a prime energy source. In sections 7.1.1 and 7.1.2 we described examples of the development by other species of tools for the acquisition of food, as well as human development of tools for the same purpose. However, this is the first case of a species domesticating another species toward the same end. The domestication of the wolf/dog is especially interesting because among the other early animal domesticates (sheep, pigs, goats and cattle), it is the lone species that was only secondarily a food. The domesticated dog evolved

to have many uses, including tracking, corralling, herding, guarding, fighting, and carrying.[54]

Like most of the early animal domesticates, plants were domesticated primarily for food. Initially, agro-forestry or forest gardening—typically along self-irrigating river courses—involved periodically burning underbrush to remove deadfall to permit new plant growth, which in turn attracted foraging herbivores. Fire scorched earth associated with animal remains dates back at least 300,000 years as noted in Section 7.2. The earliest identified starch food processing was from preparation of wild foods, like cattail and fern roots. The preparation involved cutting, peeling, drying, grinding and cooking, and was dated at 30,000 years ago.[55] This beginning overlapped with the adventitious human selection of plant species grown from seeds picked for nourishment and seeds from revisited waste dumps at former camping sites. Some of the earliest cultivation of seeds occurred at least 8,000 years ago and included barley, peas, wheat, rice, and millet.[56]

Social organization of these proto-farmers was fluid, with small itinerant bands following ancient seasonal migratory paths which aperiodically brought them together briefly in larger cooperative gatherings, e.g., large animal hunts. Toward the late Pleistocene (~100,000 years ago) the climate became more variable.[57] In addition, the growing human population and improved hunting techniques decreased the number of available hunting habitats and the availability of big game species. This forced greater reliance on less preferred foods, such as plants and smaller animals, and eventually led to planned farming: husbandry of selected animals and cultivation of selected crops.[58] Early (9,000 year ago) farming methods involved tillage with sticks and fertilization with waste and animal remains, such as manure and fish. Stream deltas and river flood plains offered the advantage of level, tillable, fertile, periodically watered land. The harvested food supported population growth and the establishment of year round settlements in spite of their initial low productivity.[59] The earliest (about 5,000 years ago), largest, and most enduring population centers were in the flood plains of large rivers like the Nile, Indus, Tigris/Euphrates, Ganges, and Yangtze.[60] The large size and long duration of these centers, and the relative sophistication of their cultures provides convincing evidence of the importance of farming to human cultural evolution. Further, their spread along East-West rather than North-South global axes implicates common seasonal and climatic factors and thus the importance of population sustenance from crop and husbandry economies. Their geographic cultural imprint continues in the dominance of Indo-European and Chinese languages today.

Reliable, storable, high energy food supported fast growing populations engaged in nonagricultural activities, such as laborers, craftsmen, tradesmen, professionals, politicians, and soldiers. The ensuing differential disparities in wealth and power reflected a degree of social stratification not possible in groups of itinerant hunter-gatherers.[61] Collective action by large groups made them formidable foes for hunter-gatherer and proto-farmer groups. Indeed the need to protect stored goods and livestock likely resulted in the development of arms technology and military prowess. In turn this combination of military technology and concentrated population may have allowed the early farming groups to take over the territory of less sophisticated hunter gatherers.[62]

In time, farming produced food surpluses, which enabled a decrease in the interval between child births, increased population density, permanent settlements, and an increase in the number of people engaged in nonagricultural work. In turn, these changes resulted in technological developments that augmented the changes initially brought by domestication. These included the development of written language, trade and accounting practices, as well as food processing, storage, and transport.

The domestication of species also produced major changes in the human genome. One of these enabled the development of immunity against infectious diseases transferred from their large domesticated animals. Thus, an advantage of farmers over non-farming peoples was their early adaptation to the epidemic diseases, such as measles, smallpox, tuberculosis, and influenza. Almost all (13 of 14) of these animals were domesticated by Eurasian peoples. Consequently, mixing with Eurasian immigrants decimated non-Eurasian populations around the world simply through contact.[63]

Other changes in the human genome brought about by domestication were the development of tolerance for wheat protein, milk sugar, and living at high altitudes. The ability to digest the protein (gluten) of wheat, oats, barley and rye (gluten tolerance) was critical to the survival of early agriculturalists who were not using maize or rice.[64] The ability to digest wheat protein in particular had critical advantages because of its relatively high protein and fiber content, its nutrient value in essential vitamins and minerals, its durability compared to meat (available all year), and the ease of baking it into edible, durable, non-crumbling cakes simply by adding water and heating. Consequently, it became a staple crop, capable of sustaining health when animal protein was in short supply. In light of these survival advantages to humans, it is not surprising that dogs have adapted to tolerate gluten as well.

Tolerance for milk sugar (lactose) is normal in infants but is usually absent in older individuals because they lose the ability to produce the enzyme lactase. Adult tolerance for lactose evolved in part because at latitudes in which ultraviolet light is inadequate to produce vitamin D, milk supplies the vitamin D needed for absorption of calcium.[65] High altitude (above 4000 m) tolerance evolved in people living in the Andes in part as a result of increased levels of oxygen-binding hemoglobin in the blood[66] and in part because of the domestication of the potato[67] and high protein grains such as quinoa.[68] On the Tibetan plateau, adaptation to altitude evolved in part by increasing the efficiency of oxygen extraction from the blood[69] and in part by the domestication of millet, and eventually of imported barley and wheat. The plant domestication at high altitudes involved adaptation to a shorter growing season, as well as increased frost and drought resistance. It is apparent that these examples of plant and animal domestication involve not only genetic changes in plants and animals but also genetic, societal and habitat changes in humans. It is likely that future human adaptations will result in the reduction of dietary preferences that had survival advantages for hunter-gatherers but now are liabilities. Such preferences and their related pathologies include preferences for sugar (type II diabetes), salt (hypertension), calories (obesity), and addictive substances (addictions).

7.6. Industrialization

The ability to produce food in excess of what was needed to sustain a single person meant that (1) some people could do things other than grow or capture food, and (2) people could move away from the immediate source of food. Freeing individuals from direct procurement of food and the need to find or follow food meant that individuals could be concentrated in one place. This can be seen in the fact that a square kilometer of land sustained less than one hunter-gatherer, from 1-3 shepherds or herdsmen, and 10-60 primitive farmers. Traditional farming through the 19th century could support roughly 100 – 1000 people per square kilometer, while modern farming techniques in the 20th century boosted that number to 800 – 2000.[70] The resulting increase in population density, and the division of labor that freedom from food production enabled, brought about the next major social innovation—the industrial revolution. While the advent of agriculture increased the energy rate density per farmer compared to that for a hunter-gatherer, an even greater impact on energy consumption was brought about by industrialization (Fig. 7.4).

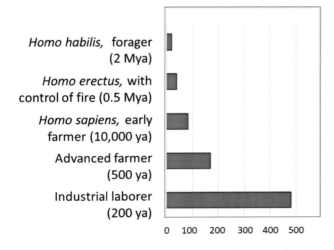

Fig. 7.4. Growth in energy consumption (energy rate densities) from 2 million to 200 years ago. *Data from Chaisson (2001).*

The industrial revolution began in Great Britain in the second half of the 18th century, then spread through Belgium and France to the rest of Europe. By the end of the 19th century, it was well established in North America, and gaining momentum in Japan, parts of Latin America, and Russia. By the start of the 21st century, industrialization was global in its influence, and largely in distribution.

The industrial revolution was driven by technological innovations that enabled a vastly increased use of natural resources and the mass production of manufactured goods. They included (1) the use of new basic materials, especially iron and steel; (2) the use of new energy sources, such as coal and petroleum, and new powers of motion, including the steam engine, electricity, and the internal-combustion engine, (3) the invention of new machines that permitted increased production with a smaller expenditure of human effort, (4) the factory system which organized work in a new way, entailing increased division of labor and specialization of function, (5) important developments in transportation and communication, including the steam locomotive, steamship, automobile, airplane, telegraph, and radio; and (6) the increasing application of science to industry.[71]

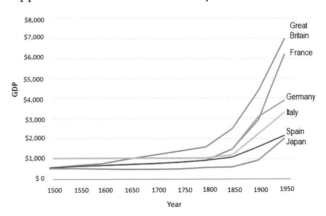

Fig. 7.5. Growth in gross domestic product per capita (GDP) in Europe and Japan from 1500 to 1950. *Based on data from Maddison (2007).*

The industrial revolution was accompanied (in some cases, made possible) by sweeping socioeconomic changes, including (1) agricultural improvements that increased food production, (2) economic changes that resulted in the decline of land as a source of wealth, a wider distribution of wealth, and increased international trade, (3) political changes reflecting the shift in economic power and the needs of an industrialized society; and (4) sweeping social changes, including the growth of cities, the development of working-class movements, and the emergence of new patterns of authority.

While many new skills were acquired by individual workers, some of the cultural consequences of industrialization were negative. Workers lost their independence since the employer owned the resources and tools of production. Freed from living on the land, cities grew into crowded, unsanitary places to live. Wages were depressed by the need for and use of unskilled labor; and the factory system led to cluttered, unsafe working places and conditions.[72]

Despite the negative social and cultural consequences of industrialization, society as a whole saw a dramatic increase in productivity and wealth (Fig. 7.5), thereby providing the resources for individuals to consume and degrade more energy. In so doing, industrialization generated a steady stream of machines capable of transforming ever increasing amounts of energy (Table 7.2).

Industrialization has accelerated the inexorable evolutionary trend toward cultural change and increasing complexity of social organization. Notwithstanding the numerous negative consequences of the industrial revolution, the pressure toward innovations that consume and degrade ever increasing amounts of energy has proved more compelling. This provides strong evidence for the pervasive operation of the PLA, in accordance with the SLT. The PLA and SLT inevitably lead toward an increase in the net complexity of society, as much as they do toward increasingly complex physical topography, chemical heterogeneity, and biological organization.

Since complexity and information are inextricably related, it follows that the more complex societies became, the greater would be their need to process larger amounts of information in a more effective way. We next turn, therefore, to a brief outline of the history of that trend.

Table 7.2. Evolution of machines and their energy consumption[73]

Introduced	Machine	Energy Rate Density (ergs/sec/g)
1888	4-stroke Otto engine	4×10^4
1900	Multi-cylinder Daimler engine	25×10^4
1978	US-manufactured automobile	59×10^4
1988	US-manufactured automobile	68×10^4
1998	US-manufactured automobile	83×10^4
1905	Wright brothers aircraft engine	1×10^6
1918	Liberty aircraft engines	7.5×10^6
1943	Whittle – von Ohain gas turbines	10×10^6
1990	Boeing 747 commercial airliner	20×10^6
1965	Phantom F-4 fighter jet	300×10^6
1995	Nighthawk F.117 fighter jet	8000×10^6

7.7 Origin of Information Technology

Biological information has been stored in the genetic machinery of the cell since the origin of life. Experiential information has been stored as memory in the nervous systems of probably all animals, and by some mechanism in some plants. But not until the rise of cognitive ability in animals with advanced nervous systems did information begin to be recorded within the lifetime of individuals, for use by that organism and other individuals. The ability of non-human animals to record symbolic information is nowhere evident, but—as with language—we may simply not know what to look for. What we know for certain is that humans began to record information at least 80,000 years ago, and developed methods for systematically processing information vital for the function of increasingly complex and energy-consuming societies with the advent of agriculture. Industrialization brought new, efficient ways of storing and processing information, leading to the computer revolution that ensued in the 20th century.

7.7.1 Writing

Systematic marks on a stone, perhaps serving as a count for some quantity, provide the oldest evidence of external information storage by humans.[74] A stone with etch marks was found in Africa, and is dated at 80,000 to 100,000 years ago. However, the systematic recording of information in the form of written language is a relatively recent technological innovation. There appears to be general agreement that the earliest writing system was devised in Sumeria, about 8,000 years ago, beginning as script and leading to cuneiform.[75]

In time, writing evolved in Europe into the alphabetic system of the Greeks. Because the alphabet represents phonemes, or units of sound in the spoken language, it is known as an alphabetical system. In China, however, a logographic form of writing evolved, in which symbols represent morphemes, which are objects or concepts, in the spoken language. In both cases, it should be noted that writing is a symbolic representation of the spoken language—not a direct symbol of an object or idea.

Alphabetical and logographic systems of writing are very different, but each has evolved to reflect the language each represents in the most efficient way possible. The symbolic representation of phonemes (sounds) by an alphabetical language enables the maximum permutations of utterances to be portrayed for a language in which words convey specific and often subtle differences in meaning. A logographic system, on the other hand, better serves to reflect languages such as Chinese and its derivatives[76] in which many morphemes (meanings) share the same or similar phonemes (sounds). Other forms of writing exist as well, such as the syllabic systems characteristic of Arabic and Hebrew. All writing systems appear to have evolved in a manner most suitable for the structure of the language which they reflect.

The history of writing has revealed a consistent trend toward representing "structural levels of spoken language . . . to construct an efficient, general, and economical writing system capable of serving a range of socially valuable functions."[77] While speech is ephemeral, writing is recorded speech that persists through time. As such, it represented a major step on the road to the technological evolution and industrialization that turned the human race into the ultimate energy consuming species on the planet.

7.7.2 Publishing

For centuries after the innovation of writing, the creation or copying of texts by hand was the only means of propagating the written word. This served as a marked advance over oral transmission of information, as it provided a record of information that survived (in principle) for an indefinite length of time. The same information could thus be accessed over multiple generations and across geographical distances. The conventions of writing, including its alphabets and iconography, however, remained a monopoly of the ruling class, which in most cases was theocratic. Only with the invention of printing, did this monopoly begin to be broken, allowing the benefits of written information to be shared by a more general populace.

The earliest printing appears to have originated in China, perhaps as early as a thousand years BCE. If so, it was not exploited on a wide scale until about the 6th century CE. The Chinese were also the first to invent movable type, in the 11th century.[78] This invention greatly increased the efficiency of printing, but was not exploited in its country of origin. The invention of the printing press in Europe, generally attributed to Johannes Guttenberg in Germany between 1440 and 1450, encompassed a whole craft of technological innovations, from movable type to ink, paper, and the duplication process. Mass reproduction of recorded information thus became possible, and proceeded to transform societies, especially after even greater efficiencies were achieved through mechanization of type setting and printing, coinciding with the spread of literacy throughout a larger segment of the population in the 18th century and thereafter.

The power of information is reflected in the vigor with which the ruling class and established religions sought to control its dissemination once printing made access to information generally available. In ancient times, it was only in those societies that were essentially non-theocratic—like Hellenistic Greece, Rome, and China—that writing was allowed to be disseminated outside the court or church. The widespread publication of all manner of ideas, first made possible by the printing press in western culture in the mid-15th century, was strongly resisted by governments threatened with the spread of anti-authoritarian ideologies until well into the 18th century, and by many religious institutions up to the present day. Because the printed word constitutes the major informational archive of a society, when one culture is overrun or superseded by another, among the items first targeted for destruction are the manuscripts of the vanquished, as carried out, for instance, by Shih Huang-ti in China in 213 BCE, the Spaniards in Mexico in 1520, and the Nazis in the 1930s.[79]

7.7.3 Calculating Machines

The advancement of science and development of technology is based in large measure on the ability to make precise quantitative calculations. This most basic form of information processing has evolved, both through conceptual innovations in the treatment of quantitative relationships (mathematics) and technological innovations that have rendered the calculations faster and less laborious.

Arithmetic must have been an invention of necessity, starting with the basic need to count the number of mouths to feed in the band, the number of hunters available, and the number of arrows required to be taken on the hunt to yield a given probability of success. More precise computations would have been required once the agricultural revolution was underway and commerce had begun. From these practical needs, an understanding of numbers and their manipulation would have provided the basic tools for the more abstract manipulations giving rise to the mathematics needed for calculating geometric relationships, astronomical events, and distances and time.

A recapitulation of the history of mathematics is beyond the scope of this book. Suffice it to say that different cultures have introduced new and more complex ways of processing quantitative information in succession.[80] While the limited evidence indicates that ancient Egypt was proficient in arithmetic calculations, those abilities were apparently directed only toward pragmatic ends. In Mesopotamia, mathematical knowledge was more advanced; basic principles of geometry and trigonometry had been introduced in Babylonia by the time of Hammurabi in the 18th century BCE, and the Persians were using math in astronomy by the 6th century BCE. Building on these origins, the Greeks raised geometry and trigonometry to new and higher levels of complexity and abstraction. Comparable advances by indigenous cultures in the Western Hemisphere can be inferred from their sophisticated astronomical calculations. Geometry was extended and algebra developed to a more sophisticated degree largely in the Arab world by the 10th century. However it would have been lost with the collapse of the Roman Empire had Islamic scholars not transmitted the Hindu-Arabic system of denoting numbers to Europe during the middle ages. In southern Asia earlier, the Chinese and Indians had developed the first decimal systems in numerology, greatly simplifying the calculation of large numbers. With the birth of modern science in the Renaissance, Europe became the center for innovation in higher mathematics—a trend that has continued almost to the present day.

As mathematics became more complex, the need for tools to help with the computation of large and complex numbers, through multiple operations, became more pressing. Thus the abacus, a board with movable counters, came into existence as the ancestral calculating machine. Probably of Babylonian origin, it was widely used in Asia, Arabia, and Europe by the middle ages.

The use of logarithms for making mathematical calculations, developed by John Napier in Scotland in 1614, provided the theoretical basis for the invention of the slide rule. Edmund Gunter (1581–1626) devised the earliest known logarithmic rule as an aid for nautical calculations by sailors. William Oughtred, another Englishman, designed the first adjustable logarithmic rule, but the familiar inner sliding rule was invented by the English instrument-maker Robert Bissaker in 1654. With upgrades, such as the "log-log" slide rule for calculating powers and roots of numbers introduced by Peter Roget (of Roget's Thesaurus) in 1814, the slide rule became an indispensable instrument for scientists and engineers until it was supplanted by hand-held electronic calculators in the late 20th century.[81]

While the slide rule provided a convenient, hand-held apparatus for making quick calculations of limited precision, actual machines with moving parts were being developed in parallel. Blaise Pascal invented a digital arithmetic machine in 1642, and Gottfried Leibniz created a more advanced version later in the same century. Calculating machines that were progressively smaller and easier to use were introduced in succession, leading to desktop machines for quickly and easily doing long strings of calculations by the early 20th century. By the late 20th century, solid-state electronic devices had ushered in new miniaturized calculators for pocket or desk top that could perform simple mathematical functions, including normal and inverse trigonometric functions, as well as basic arithmetical operations. Endowed with an increased capacity for information storage and programmed execution, these sophisticated calculators were able to employ interchangeable preprogrammed software modules capable of 5,000 or more program steps. Meanwhile, a full-fledged revolution in larger, more complex, and dramatically more sophisticated, computers was underway.

7.7.4 The Computer Revolution

A calculating machine that can be programmed to carry out different types of computations is a computer. In 1805, a French weaver, Joseph-Marie Jacquard, devised a way to instruct his loom to create different patterns in woven fabric. The placement of variably colored threads was controlled by the motion of rods whose position was determined by a pattern of holes punched in cards that could be inserted interchangeably into the machine. Thus, the task of weaving was changed from a labor-intensive chore to a mechanical task directed by a spatial pattern (an informational program). This can be said to be the first programmable machine.[82]

In 1833, Charles Babbage, an English mathematician and inventor, designed the first machine (the Analytical Engine) with all the components of a modern computer. They included a "mill" (central processing unit, in today's vocabulary), a memory storage unit, input reader, and output printer. The Analytical Engine was thus the first general-purpose, fully program-controlled, automatic mechanical digital computer.

By the start of the 20th century, mechanical calculators and typing machines had transformed the process of handling large amounts of information from a purely manual to a machine assisted operation. As the century unfolded, the pace of technological evolution accelerated through a succession of innovations that increased information processing capability in an exponential fashion. Those innovations were, in order, electromechanical switches, relay switches, vacuum tubes, transistors, and integrated circuits. Each technological advance increased the proficiency of computers—their ability to process a given amount of information per unit of time and cost—by at least an order of magnitude (Fig. 7.5). The following list of watershed events in the history of computer development is based on the review by Freiberger et al. (2014).

As electromechanical switches used in calculators were giving way to relay switches, then to vacuum tubes in the late 1930s, two different computers emerged at the Massachusetts Institute of Technology (MIT) and Harvard University, respectively. Vannevar Bush at MIT developed the first modern analog computer, and Howard Aiken at Harvard, in collaboration with International Business Machines (IBM), developed the first fully functional digital computer, known as the Harvard Mark I. A huge machine, consisting of about 750,000 separate parts, it weighed five tons and was more than 15 meters long. Meanwhile, George Stibitz at Bell Laboratories created the first computer network, linking a series of machines through telephone lines to serve more than one user.

In the early 1950s, John Backus at IBM created FORTRAN, a programming language and a compiler (for converting intuitive language into information readable by the machine) that would produce code that ran virtually as fast as hand-coded machine language. This vastly increased the ease of the program-writing process, and continued to be the favored programming language until the development of BASIC in the early 1970s by Microsoft.

In 1957, the Digital Equipment Corporation (DEC) introduced the first integrated circuit—a set of interconnected transistors and resistors on a single silicon wafer, or chip, selling initially for around $20,000 (falling to $3,000 by the late 1970s). This made possible the first true mass-market minicomputers, as stripped-down versions of mainframes that could be operated without specialized computer knowledge and could sit on a desktop or lab bench. When Don Lancaster in 1973 proposed the first television typewriter, a machine that linked to a television screen to display typed-in alphanumeric characters, the computer monitor was born. With this machine, called the Altair, the prospect of personal computing on desktops fully emerged.

Simultaneously, Bill Gates and Paul Allen were starting to call their partnership Microsoft, and were developing a version of the BASIC programming language that could run on the Altair. In 1976, Commodore started selling the first microcomputer; and the Tandy Corporation through its RadioShack outlets began mass distribution of the TRS80 the following year. Microsoft went on to develop versions of BASIC for nearly every computer that was released.

Motivated to make computing easier, personal, and more intuitive, Stephen G. Wozniak and Steven P. Jobs invented their first Apple computer in 1976. They recruited A. C. ("Mike") Markkula, a retired semiconductor company executive, to help them write a business plan, line up credit from a bank, and hire an experienced businessman to run the venture. Using conventional business and marketing techniques, Apple Incorporated started selling its wildly successful Apple II personal computer in 1977.

In 1979, Dan Bricklin and Bob Frankston developed VisiCalc, an easy-to-use electronic spreadsheet that suggested the business and commercial utility of the smaller personal computers. Long a leader in the development of calculating machines and mainframe computers for business purposes, IBM entered the market with its version of a more powerful "personal computer," the IBM-PC, geared toward business and scientific applications. In partnership with Microsoft, IBM introduced the MS-DOS operating system which could be used by any machine emulating the IBM-PC architecture. This spawned the growth of a whole host of clones, the first major one being the Compaq computer which began sales in 1982.

That same year, Apple introduced the Lisa computer, which made use of a graphical user interface (GUI), pioneered by Xerox Corporation. The GUI replaced the typed command lines of previous computers with graphical icons on the screen that invoked actions when pointed to by a handheld pointing device (soon called the "mouse"). The Lisa was expensive and not a commercial success, but a slimmed down version called the Macintosh was introduced in 1984. It was immediately successful, and led the way toward an irreversible transition to the use of GUIs as the interface of choice between users and computers. The Microsoft version (Windows) accelerated and consolidated the trend.

As Fig. 7.6 shows, the speed of computers accelerated at a rate greater than exponential throughout the 20th century. Note further that the change in rate (computations/second) is normalized to cost, which itself is a proxy for materials and energy. Thus, the power of computers to process information, whether measured by speed or by units of energy and mass required, has increased at an astounding rate. Compared to biological evolution—itself an acceleration over the rate of change in the physical world—technological evolution has become the dominant driver of change in human society and culture. The PLA predicts that change will flow faster and more completely as pathways for doing so are opened up by technological innovation. And in so doing, more total energy will be degraded in accordance with the SLT, in the process of concentrating and processing more energy (decreasing entropy) at the local level while the entropy of the universe as a whole continues to grow.

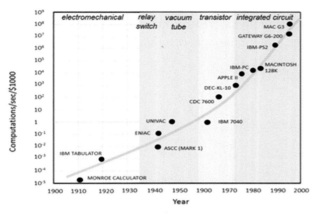

Fig. 7.6. Evolution of computer proficiency during the 20th century. Shaded areas indicate the approximate time spans dominated by indicated technology. The computers shown are only a small fraction of those actually produced. *Based on data from Kurzweil (2005).*

7.7.5 Information Everywhere

The miniaturization made possible by integrated circuits spawned the advent of even smaller, hand-held computers, called personal data assistants (PDAs), such as the Palm Pilot introduced in 1996 by the Palm Computing Company. With personal computers emerging as essential productivity tools in the late 20th century, other innovations were making information both more ubiquitous and more personal. By the 1960s, semiconductor technology made it possible for information to be processed in rapid succession for multiple users, alternating among users so fast that the computer appeared to be dedicated to the task of a single user despite the access of the system by many others simultaneously. This led to the development of networks of interconnected terminals linked through host-computers (or simply "hosts") for servicing a large number of users at the same time.

The Advanced Research Projects Agency (ARPA) of the U.S. Department of Defense established the first host-to-host network connection on Oct. 29, 1969.[83] This network connected time-sharing computers at government-supported research sites, principally universities in the United States. Known

as ARPNET, it was one of the first general-purpose computer networks.

In 1980, the National Science Foundation (NSF) funded construction and the initial maintenance of a network supplementary to the ARPNET, known as the Computer Science Network (CSNET). During 1985-86, NSF funded establishment of five supercomputing centers—at Princeton University, the University of Pittsburgh, the University of California, San Diego, the University of Illinois, and Cornell University. This formed the backbone of the NSFNET, an interconnected system that provided access to the entire scientific and academic community.

The need to connect networks in the United States with those in Europe led to the Internet. A truly global network was made possible by adoption of an open architecture system with defined standard interfaces (the TCP/IP standard), which allowed different computers with various operating systems and different networks to communicate with one another.[84] This capability was accelerated and made easily accessible to everyone when Tim Berners-Lee of England and others at the European Organization for Nuclear Research (CERN) developed a hypertext protocol to make information distribution easier across all platforms and operating systems. These developments culminated in the creation of the World Wide Web. This was soon followed by development of a program called a browser at the U.S. National Center for Supercomputing Applications in Urbana, Illinois that made it easier to use the World Wide Web. A spin-off company named Netscape was founded to commercialize the technology. Other companies followed with their own browsers, and information access became truly global.[85] By 2010, it was estimated that half the world's population had access to the Internet at some level.[86]

While information technology was becoming globalized, it was also becoming increasingly personalized. Mobile telephone devices small enough for personal transport began to appear in the 1980s, relying on transmission of radio waves. An advanced mobile phone system, developed primarily by AT&T and Motorola, was publicly introduced in Chicago in 1983, and by the end of the first year of service, had a total of 200,000 subscribers throughout the United States; five years later there were more than 2,000,000.[87]

With the number of users skyrocketing, strategies were developed to handle the increase in electronic traffic. Basically, service areas were fragmented into roughly rectangular geographical units called cells, which could relay signals from one to the other, eventually connecting a sender and receiver from any point in the network. The communication devices were thus known as cellular phones, or simply "cell phones." The analog systems of the first cell phones constituted the "first-generation" (or 1G) systems, which served to send and receive voice signals like conventional land-line phones. The digital systems that began to appear in the late 1980s and early '90s were tagged as "second generation" (2G) systems. The various enhancements made possible by digital technology enabled applications such as Internet browsing, two-way text messaging, still-image transmission, and mobile access by personal computers—thus making them "smart phones." Beginning in 1985, "third-generation" (3G) cellular standards were adopted. The higher data rates of 3G systems made transfer of higher-density information, including full-motion video transmission, image transmission, location-awareness services (through the use of global positioning system technology), and high-rate data transmission, possible. Currently, cellular standards are advancing into "fifth-generation" (5G) networks.

In 2007, Apple redefined the smartphone with its introduction of the iPhone. This was a cell phone with a touch-screen interface more advanced than the graphical user interface used on personal computers, and with greater storage capacity than that of computers from just a few years earlier. The capabilities of the iPhone were soon matched by Sony Corporation's Android system. By the middle of the second decade of the 21st century, these devices had brought hand-held access to information from almost anywhere in the world, available to the majority of the world's population.

The rapid emergence of access to information both globally, through the World Wide Web, and personally, through the rise of cell phone users, over the last two decades of the 20th century, is shown in Fig. 7.7. Thus, as the 20th century came to an end, computers had become pervasive, personal, powerful, and affordable. While the physical world continues to change, and all the other species are still evolving at a biological pace, for humans, the processing of information has come to be the commodity of greatest value, hence the trajectory of the evolutionary process most salient to our survival.

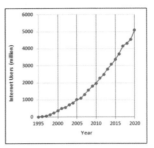 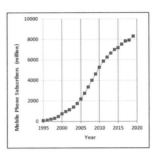

Fig. 7.7. Growth in number of worldwide (a) internet users and (b) mobile phone subscribers from 1995 to 2020.

7.8 Consequences of Technology

Technology has profoundly affected the social organization and behavior of humans—often in ways that have circled back to accelerate the evolution of technology itself.

As long as humans subsisted by gathering and hunting for their food, they were little distinguished from other animals of equivalent size and cognitive ability in their contribution to and impact upon the ecosystem of which they were a part. The domestication of animals and advent of agriculture, however, ushered in changes that greatly improved the efficiency of human effort. Farming and herding increased the per capita production of food, enabling a single human to produce more food than the individual required. As tools for tilling became more effective, and draft animals assumed more of the burden of arduous labor, plant food could be produced in greater volume per unit of human effort. Herding larger mammals with the help of domesticated dogs and horses ensured a more easily controlled and reliable supply of meat. The creation of hooks and nets facilitated fishing. Thus, a subgroup of humans could feed the entire group, freeing those not engaged in food production for other pursuits, including the further development of technology.

The division of labor made possible by more efficient food production enabled specialists, such as craftsmen, laborers, artists, and servants to assume specific roles. Increased freedom from time devoted to procuring food made more time available for storytelling, thinking, and studying—augmented eventually through writing, illustrating, computing, and publication. As social roles differentiated, the need for social organization became more acute, and societies evolved from the egalitarian structures characteristic of hunting and gathering societies, to the more complex and hierarchical social structures needed for organizing multiple tasks within an overall group effort. At the top of the hierarchy were those who controlled territory and the means of production, including the efforts of other members of the group. Thus, obtaining and controlling territory, and capturing other humans for labor and servitude, became ends in themselves, leading to social structures and behaviors supportive of human competition and aggression.

Another consequence of more efficient food production was a rise in human density. As humans came to live together in larger numbers, the social fabric changed to accommodate larger group sizes devoted to concerted objectives. The utility of warfare as an extreme form of intergroup competition was likely driven by its success in promoting overall group survival, by increasing the availability of prime agricultural land, food resources, and slaves. Human success in large-scale group warfare would have placed a premium on technological developments such as improved weaponry and the sophistication of within-group communication and tactics. It is also apparent that the adoption of an agricultural life style would have greatly potentiated population density, and so augmented putative warfare effects.

Mechanization, leading to industrialization and automation, further reduced the amount of human effort needed to manufacture goods and consume energy. Physical barriers to human mobility have been reduced, travel has become more extensive, and communication in the present day has become global and virtually instantaneous. While the earlier shift from hunter-gatherer to agricultural societies led to a plethora of culturally distinct populations largely isolated by distance and physical barriers around the world, the opposite has occurred in modern times, as technology and communication have turned all the world's societies into an interconnected and inter-communicating web of global communities.

Information technology has enabled the storage and recall of historical records with precision, erasing temporal barriers to knowledge and understanding of past events. The combination of extensive data bases and the speed of computers now provides an analytical capacity and predictive capability unknown to Paleolithic people. With all the world's populations becoming interconnected through trade and communication, the speed and efficiency of commercial transactions have transformed society into a global economy. Information has overtaken energy as the commodity of highest value in all the industrialized world.

Technological progress clearly has negative as well as positive effects. Vulnerability to disease has increased with animal domestication and urban crowding; and while antibiotics have provided unprecedented protection against some diseases, their overuse has given rise to more rapid evolution of resistant pathogens. Some deleterious medical conditions have been accentuated in the industrial world, such as some forms of cancer, decreased human fertility, and probably a number of psychogenic disorders including autism. With information collection and sharing on a global scale, ownership of information has become less secure, personal information has become less private, and political oversight and control of individual and in-group daily life have become greater in many states and governments, and of great potential abuse by rulers and governments everywhere.

At the same time, energy consumption has never been greater, and its degradation as required by the SLT is changing the biosphere at an unprecedented rate. While levels of carbon dioxide (CO_2) have been higher on Earth at times in the geological past, non-renewable fossil fuels are being extracted in ever-growing volumes, and their conversion to energy in the degraded form of CO_2 as well as their escape from production facilities as methane, are adding those compounds to the atmosphere at the highest rate in the history of the planet. That, in turn, is speeding the rate of global warming.

Industrial pollution (another manifestation of the SLT), extensive deforestation to increase agricultural land, and depletion of natural populations such as fish for food, are transforming the biosphere at an alarming rate.

All of these changes are occurring in the way they are because they are impelled to do so. Indeed, the acceleration of change in the physical and biological environment, and ultimately in society and culture, are predicted and explicable in terms of the concerted imperative of the PLA and the SLT. Because energy will be degraded as completely and efficiently as it can (the PLA) to a less useful and more dispersed form (the SLT), it inevitably must be. The implications of the action of these fundamental laws of nature, and the changes they are bringing about for the future of the Earth and its biosphere will be taken up in the next chapter.

7.9 Cooperation

Many species have evolved cooperative behavior to maximize the efficiency with which they obtain, utilize, and defend their resources. Food collection is obviously fundamental to an organism's ability to acquire and consume energy. Not surprisingly, then, species across phyla have evolved cooperative behaviors that optimize foraging for food. Group foraging and caching has developed in many species of organisms. Individuals forage and cache more efficiently as group members than as lone individuals. The best examples are caching and food storage by insects, birds, and mammals. Ant colonies have also developed living stores of food to supplant foraged stores, e.g., fungus growing and aphid milking,[89] and prudent slavery of other ant species.[90]

In many species, food collection and storage carried out cooperatively with others is based on kinship. Parental feeding of the young and teaching them to forage are ubiquitous examples of this. Such cooperative behaviors are also observed between conspecific individuals which are not kin. This is remarkably illustrated in eusocial animals which give up their ability to procreate in order to maintain the group's nest and food supply. This is also seen in group seasonal migrations of many species which

provide increased efficiencies in reducing the energy demands of migration.

The suite of behaviors that promote acquisition and protection of resources has been found to have a neurological basis, mediated by neuroendocrine mechanisms and reinforced by the neurochemical reward circuitry of the brain. Among the most important mediators of these behaviors in vertebrates is the neuropeptide, oxytocin, which is produced in the hypothalamus of the brain and secreted from the posterior pituitary gland.[91] Oxytocin-like genes and peptides have changed over the course of evolution from a strictly pro-copulatory role in ancestral marine vertebrates to a much broader social role in living mammals. Indeed in modern birds and mammals, oxytocin has general prosocial properties that increase the sensitivity of reward circuits related to pair bonding as well as social behavior facilitating empathy, affiliation, and social memory, while reducing anxiety.[92] The necessity for food consumption and the survival value of expending energy to obtain, sequester, and guard it, are reinforced by the dopamine reward system in the brain. Subjectively feeling good about performing these behaviors is congruent with their genetic programing by natural selection and their mandate by the SLT and PLA. In fact, foraging is often so effective that it leads to over exploitation of natural resources and population crash,[93] illustrating that short-term thermodynamic advantages can override the longer-acting effects of natural selection.

The willingness of humans to cooperate with others and to assist others who are non-kin is striking and demands explanation. Human prosocial behavior is quite unusual, even compared to other primates. It is not shared with our closest relatives the chimpanzees. Assistance is often offered without gain for the donor, directed toward non-kin as well as kin, and is often prosocial — that is, unsolicited.[94] A number of hypotheses have been suggested as critical to its evolution. These include extraordinary cognitive capabilities (usually assessed by the degree of brain development),[95] reduced in-group aggression,[96] social bonding,[97] dependence on cooperative efforts in foraging,[98] and extended care of the young.[99] A comparative analysis of 15 primate species clearly pointed to care of the young, initially by non-kin females, as the only factor clearly predicting proactive sociality among non-kin.[100] Interestingly, species of new world monkeys—tamarins, squirrel monkeys, saki, and marmosets—were more like humans in their social behavior than chimpanzees. These species are diurnal, arboreal, and omnivorous, live in groups, and share in the care of the young. It seems plausible to suppose that human sociality between members of groups of multiply mated pairs was the ancestral characteristic. This could have later generalized to all in-group members. Eventually, this could have been generalized to support strong cooperative bonds between members of large anonymous groupings. However it evolved, such hyper-cooperative behavior made possible human technological changes with attendant massive increases in energy consumption and information processing.

Some of the species most successful at exploiting their environments carry sociality to the extreme of eusociality, giving up their individual procreative rights to enhance the efficiency of the group. Eusociality in ants and termites is a prime example, in which reproduction is restricted to a single queen, while males and non-reproductive females have become morphologically differentiated. They have achieved ecological dominance over solitary and pre-eusocial competitors, due to the altruistic behavior among nest mates and their ability to coordinate action through pheromonal communication.[101] An example among mammals is the naked mole-rat, in which reproduction in the colony is restricted to a single queen female and several males, while all others are reproductively suppressed subordinates.[102] Such self-sacrifice pits the ultimate goal of effective energy transformation and gene pool success against individual reproduction. The fact that the occurrence of eusociality does not depend on kinship in many species suggests that eusociality is the result of the advantages of assuring the integrity of the nest and young, and the food supply, rather than the result of kinship. Eusociality has been rare in evolution; the environmental pressures sufficient for tipping the balance among countervailing forces in favor of group selection apparently being rare. That it has happened on occasion, however, illustrates the extent to which thermodynamic (energy acquisition and consumption) considerations can take over natural selection.

Group communication increases the effectiveness of social foraging in humans. Cooperation gives each group member a large survival advantage in acquiring and controlling resources. Historically recurrent examples in human culture include the clearing of land with fire to increase the abundance of wildlife, the domestication of animals, and the storage of harvested crops.[103] While hunter-gatherer societies are typically egalitarian, societies which store food are prone to social inequality,[104] illustrating that effective resource management acquires greater importance than independence and egalitarianism in the course of social evolution.

A distinguishing feature of humans compared to other species is the ability to collaborate in joint undertakings in very large groups in spite of not being eusocial.[105] Large group cooperation among humans presents unique challenges, including development and use of a shared form of communication, and the need to select, agree upon, pursue, and maintain shared goals, in spite of the heterogeneity of personal needs and the tendency to free ride.[106] The selection scenarios thought to contribute to the appearance and maintenance of large group collaboration include the hunter/forager model[107] and the intergroup warfare model.[108] Researchers in the area think both scenarios are relevant but they differ in giving one more emphasis than the other. Evolutionary mathematical modelling has recently strongly supported the warfare model with hunting/foraging playing a subsidiary role in the evolution of large group collaboration.[109] An important difference between hunting/foraging and warfare is that the demands and likely success of the latter are more sensitive to human population density and the size of the collaborating group, over a much larger range of group sizes. However this military superiority would be an advantage only if there were a resource worth acquiring and protecting.

A possible example of such an advantage occurred at a critical juncture in the evolution of *H. sapiens*. In the late Pleistocene, glaciation precipitated a wide spread period of food scarcity. The rich coastal shellfish beds of South East Africa then became a resource worth protecting and fighting for.[110] Archeological evidence indicates the shellfish beds required relatively little labor and so supported large groups of *H. sapiens* for a prolonged period. Archeological findings also reveal the concurrent evidence for the use of advanced projectile technology in the area, including microlith tipped spears, darts, the atlatl, and the bow and arrow. Beginning around 70,000 years ago, some members of this population migrated north out of Africa, possibly hastening the extinction of megafauna, Neanderthals, and Denisovans in Europe and Asia. As the sole surviving hominin species, these people eventually invaded the Arctic, the Australian continent, Polynesia, and the Americas.

What were the underlying changes in modern humans that made large group cooperation possible? Our chimpanzee relatives are able to function cooperatively in groups of from 10 to 50 animals.[111] Humans typically top out at about 150 individuals as a maximum number for one person's circle of active acquaintances.[112] Such groups are often family groups, small groups of traders, local communities and small military units. They can cooperate effectively based on personal acquaintance and frequent exchanges of information (gossip).[113] However the same attributes do not effectively support human cooperation of much larger groups. Certainly, vigorous hypersociality or prosociality beyond the reproductive bonding seen in other species was a critical development for effective in-group cooperation in *H. sapiens*.[114]

But distinctive cognitive capabilities played a role as well. Cultures of human groups uniquely share not only common perceptions of present and past events, but also of purely conjectural future events.[115] These expectations or beliefs about the future are often at odds for different groups. Because they are fictions, they are not demonstrable. Yet as a basis for assigning value, defining goals and plotting a course for future action, they are critical for insuring effective in-group cooperation and harmony. This is true whether we are considering the tribal organization of a massive seasonal bison hunt, a nation going to war, corporate planning for the worldwide marketing of a new product, or a church's evangelical mission. Modern international currency systems, financial agreements between borrower and lender, between investor and innovator, between employer and employee all ultimately require trust. At some point each participant must accept the group's vision and mission—its version of what is not demonstrable. Indeed the point has often been made that most of the world's great religions include counterfactual elements, such as supernatural beings, virgin births, etc. These beliefs test the strength of the believer's faith and serve as a barrier to facile conversion. Critically, for securing allegiance of converts to large groups, the process is readily accomplished once the leap of faith is made, coupled as it is with the cognitive in-group biases of *H. sapiens* (Chapter 8).

From the above it should be apparent that we consider the informational component of large group functioning as absolutely foundational to humans' unique energy dissipating attributes and to our future. We will return to this subject in Chapter 9 when we discuss the cognitive implications for human fertility rates of the adage that perception is reality.

7.10 Morality

Moral standards have coalesced over the millennia from proto-feelings about caring, fairness, and sharing of energy resources (sometimes to an altruistic degree), reinforcing behaviors consistent with the most effective uses of energy. Cross cultural studies[116] and studies in young children[117] indicate that some components of moral feeling and judgment may be genetically based and may have antecedents in other species.

Cheating and mechanisms to control cheating by individual group members have been evident at every stage of the evolution of the interdependence of individuals.[118] The presence of cheaters poses a special threat to societies dependent on caring and sharing of goods. If cheating reduces net group energy dissipation through unwarranted sequestration of resources, the repeated appearance of cheaters would thwart the consumption of energy contrary to the PLA. However there is recent evidence that the presence of some cheaters in a population increases the group's total energy consumption as a result of maximizing resource accumulation, population growth, and fitness.[119] While moral condemnation of cheating may persist, it may in selected instances actually promote energy consumption.

To the extent that moral imperatives are concerned with what *ought* to be, a fair question to ask is whether the evolution of moral behavior (seemingly geared toward promoting the success of the group gene pool) sheds light on the *is-ought* issue raised by David Hume, who suggested that there is no obvious way to deduce prescriptive statements about what *ought* to be from descriptive statements of what *is*.[120] As noted above, the evolution of moral behavior has strong support in the ethological literature—from control of cheating, group cooperation, interspecies cooperation, eusociality, and human ontogenetic development. The evidence is clear that certain tendencies have been programed by evolution into the genetic makeup of each species, to promote its ability to survive and reproduce. This process manifests itself in humans in the form of certain intrinsic biases and attitudes. For example, the fundamental perception of what is "fair" and the attendant dislike of cheating have been observed in very young children and suggest an unlearned, heuristic origin.[121] If one assumes that values and attitudes (judgments about what *ought* to be) have their anthropological origins in the evolution of tendencies that benefit the long-term success of the population gene pool—not the least of which is the efficient use of energy—then one could argue that the reality of natural selection (a fact of nature) does lead to prescriptive behavior that benefits the group (what ought to be), thereby serving as a link between a natural phenomenon and an ethical value.

The relative merit of ethical *ought* judgments about an act being judged "good" or "bad," can be assessed by an estimation of the likelihood that commission of the act will bring about the desired goal.[122] The desirability of the goal itself, in turn, would be context dependent—on the particular energy source, time frame, and group identity of beneficiaries. The latter two are critical because they define who we share our energy resources with and when. Indeed, our phylogenetically ancient antipathy for individuals identified as

belonging to the out-group has been hypothesized to be fundamental to our moral perception of what "bad" is. Conversely, in-group members and their behaviors help define what is "good."[123] Again, we see normative attitudes *within the group* arising out of the natural tendency to promote the survival of the group—a descriptive fact attributable to natural selection.

For individual decision making, the significance of the present model is that guarding against self-interest becomes a struggle to bend our natural proclivity to competitively consume material wealth toward a more benign form of energy consumption. Some may find the directive expressed in the SLT unacceptable as a foundation for human behavior. However, it is the *de facto* direction of life processes. It controls our short-term actions and dictates our long-term interests. If we choose to ignore it and act as if it is not relevant, it will eventually lead to the collapse of cultures and the extinction of the human species.

Mortal self-sacrifice pits the ultimate goal of effective energy transformation against the proximate evolutionary goals of individual survival and reproduction, in the starkest possible terms. There are many examples of individual self-sacrifice in human warfare,[124] which is accorded the highest moral value in most societies. This is true whether or not the sacrificing individuals are genetically related to the host group. Self-sacrifice is an altruistic behavior that promotes survival of the group, irrespective of genetic relationships.

Over time, successive religions have fostered cooperation in ever larger portions of humanity, spanning successively boundaries of kinship, geography, race and language. In this context, for human culture, morals embodied in codes of ethics, in laws, and in values expressed in social norms are agreements on the regulation of behavior in groups to control cheating and to ensure a semblance of balance between advantages in energy control for the individual and the group. Thus the expression of the SLT and PLA in the natural and biological realms have homologies in animal and human behavior and in human culture. Neither the SLT nor the PLA "cares about" kinship. This clearly dissociates the ultimate and proximate causes of evolution.

The development of a scientific calculus of ethical behavior is ongoing. Here we are suggesting simply that the function of ethical systems in human societies can be viewed through the framework of how the SLT and PLA operate in negotiating between the interests of the individual and the group. Aggregating the effects of energy transduction across individuals to include any sized population has many exemplars: metabolism as energy consumption at several levels in the ecosystem;[125] energy consumption from the cellular to the behavioral level;[126] the subjective sense of well-being in individuals and groups;[127] factors contributing to happiness including compatibility with one's partner, values and priorities, religion, working hours versus leisure hours, social participation, and healthy lifestyle;[128] economic measures for a composite national or corporate assessment of happiness of a population;[129] and measures of social justice and welfare economics.[130]

Summary of Chapter 7

Humans have come to dominate the environment like no other species in the history of life, by consuming and degrading energy at an unprecedented rate. The only surviving Hominin, *Homo sapiens*, emerged from a host of competing species in the late Pleistocene by mastering the use of tools, gaining control over fire, acquiring the ability to communicate verbally with symbolic language, and domesticating plants and animals. In the process, humans changed from bands of hunters and gatherers to agrarian societies, then urbanized and stratified civilizations that promoted group effort, identity, and cohesion through shared cultural values expressed in art, music, dance, ceremony, and literature. A succession of technological innovations, including the ability to publish, manufacture, calculate, compute, and communicate on a global scale, elevated the recording, manipulation, and analysis of information to the commodity of greatest value in the perpetuation of the species.

The use of tools by humans began as an extension of abilities shown by many other mammals and birds, and by some insects to a degree. Humans leveraged additional characteristics, including the free use of both hands, a bone and muscle structure well suited for grasping, freely rotating shoulder joints, exquisite eye-hand coordination and fine digital motor control. Eventually, tool use was augmented and accelerated by development of the cognitive capacity to envision uses and foresee design alternatives, to persuade large groups of others of the veracity of these alternatives, and to manufacture tools of increasing sophistication and utility. The consequence of tool use, whatever its proximate purpose, was to increase the effective use of energy.

The control of fire and the evolution of language contributed both directly and indirectly to energy transformation. The mastery of fire contributed to energy consumption directly through increased burning of plant products and fossil fuels, and indirectly through the social cohesion engendered by the campfire, the increased caloric and nutritional benefit of cooked food that fed an ever-enlarging brain, enhanced social communication, and division of labor. The evolution of human language provided a basis for the accumulation, storage, and transmission of information—an acknowledged form of energy dissipation. Language—initially gestural—was overlain and eventually superseded by spoken language and subsequently reinforced by written language. The opportunities for group energy expenditure were greatly enhanced by the ability to describe complex conditional contingencies between imagined events. Language permitted the conscious review, reconstruction, and planning of experiences and the sharing of these experiences with others. Culture evolved in tandem with fire and language as fundamentally a stylized use of information for promoting group identity and cohesion. Much of this group activity was mediated by artistic performances (dancing, singing, decoration, feasting, drinking and storytelling) which served many social functions: aiding in recall of collective history, instruction of children, social bonding, and arousing the group to collective action in securing the energetic needs of the larger group.

Humans are not the only species to make use of cooperative behavior, but probably extend cooperation to a level and sophistication not seen in any other living organism. The selective advantages of cooperative behavior by humans, which were neither the largest nor strongest nor fastest member of their ecosystem, very likely interacted with the selective advantages of language and the cognitive development of foresight.

Morality, though often thought of as a uniquely human trait by those unaware of observations in other species suggestive of ethical dimensions to their behavior, can also be ultimately tied in many ways to an evolutionary consequence of bending behavior toward what is best for the survival of the group, and therefore for perpetuating the population gene pool.

As technology has developed, including the ability to store, process, and communicate ever increasing amounts of information, the impact of mechanization along with the burgeoning of the human population, has begun to affect and largely degrade the environment faster than the natural cycles of change can absorb. In an age where the role of information has overtaken the consumption of energy as the dominant endpoint of the evolutionary process, the sustainability of the biosphere has come into question. Because of the overwhelming impact of our technology and energy consumption, the fate of our planet now rests in our hands — in the intelligence with which we use or misuse our tremendous power to consume and manipulate both energy and information.

References and Notes

[1] Schulze-Makuch & Irwin, 2008

[2] Map of Life, 2012

[3] Rutz et al., 2010

[4] Ruxton & Stevens, 2015

[5] Semaw et al., 1997; McPherron et al., 2010; Ward et al., 2014

[6] Wood, 1996

[7] Movius, Braidwood & Kuiper, 2014

[8] Hartenberg, 2014

[9] Chang, 2014

[10] Connan, 1999

[11] Henry, Brooks & Piperno, 2011; Roebroeks & Villa, 2011

[12] Goren-Inbar et al., 2004

[13] Wrangham & Conklin-Brittain, 2003

[14] Wobber, Hare & Wrangham, 2008

[15] Eastwood, 2003; Wrangham & Conklin-Brittain, 2003; Carmody & Wrangham, 2009; Carmody, Weintraub & Wrangham, 2011; Groopman, Carmody & Wrangham, 2015

[16] Martin et al., 1985; Aiello & Wheeler, 1995; McBrearty & Brooks, 2000; Hladik & Pasquet, 2002; Fish & Lockwood, 2003

[17] Wrangham et al., 2009; Baumeister & Masicampo, 2010

[18] Dunbar, 1993

[19] Wrangham & Conklin-Brittain, 2003

[20] Wobber, Hare & Wrangham, 2008

[21] Barrett & Arno, 1982; Orton, Tyers & Vince, 1993; Bellomo, 1994; Budd & Taylor, 1995; Goren-Inbar et al., 2004; Clark & Harris, 2005; Naveh, 2006; Henderson, 2007

[22] Neubauer, 2012

[23] Arbib, 2008; Tomasello, 2008

[24] Pollick & de Waal, 2007

[25] Gentilucci & Corballis, 2006

[26] Corballis, 2004

[27] Meguerditchian et al., 2012

[28] Smaers et al., 2013

[29] Brown, Martinez, & Parsons, 2006; Gentilucci & Corballis 2006

[30] Foundas et al., 1998

[31] Rizzolatti & Arbib, 1998

[32] Rizzolatti & Arbib, 1998; Arbib, 2008; Arbib, Liebal & Pika, 2008

[33] As stated by Cowen 1995, "The cooperation required to build, start, control, maintain, and transport a fire is very high. It's difficult . . . to imagine a campfire without conversation."

[34] Wood & Eagly, 2002; Reiches et al., 2009

[35] Kaplan et al., 2007

[36] Harris, 1978

[37] Axlerod & Hamilton, 1981; Wright, 2000; Eckhardt, 2006; Scott-Phillips, 2007

[38] Ridley, 2010

[39] Baumeister & Masicampo, 2010

[40] Elkin, 1964; Pfeiffer, 1985; Dissanayake, 1988; Boyer & Lienard, 2006

[41] Whitehouse, 2004

[42] Harris, 1978

[43] Karnani, Pääkkönen & Annila, 2009

[44] Baumeister & Masicampo, 2010

[45] Laski, 1961; Gardner, 1973; Santos Granero, 1991

[46] Csikszentmihalyi, 1991; Seligman, 2002; Baumeister et al., 2007a

[47] Koob, 1992; Knutson et al., 2001

[48] Esposito et al., 1984

[49] Shipman, 2014

[50] Flannery, 1969; Helmer et al., 2005

[51] Shipman, 2014

[52] Thalmann et al., 2013

[53] Finlayson, 2009

[54] Arnold, 1979

[55] Revedin et al., 2010

[56] Zeder et al., 2010

[57] deMenocal, 2014

[58] Flannery, 1969; Diamond, 2002

[59] Bowles, 2011

[60] Diamond, 2002

[61] Testart, 1982

[62] Bowles, 2011

[63] Diamond, 2002

[64] Greco, 1997

[65] Flatz & Rotthauwe, 1973

[66] Storz, 2010

[67] Pickersgill & Heiser, 1978

[68] Galwey, 1995

[69] Storz, 2010

[70] Neubauer, 2012

[71] Daggett, 2019

[72] Editors, 2014 "Factory System"

[73] Based on data from Chaisson, 2001

[74] Tattersal, 2014

[75] Olson, 2014

[76] Here used generically to mean Mandarin, Cantonese, and other dialects in China, Korea, Japan, and Southeast Asia

[77] Olson, 2014

[78] Unwin, 2014

[79] Ibid.

[80] Berggren & Knorr, 2014

[81] Editors, 2013, "Slide rule"

[82] Freiberger, Swaine & Pottenger, 2014

[83] Dennis, 2014, "Internet"

[84] Dennis, 2014, "Internet"

[85] Freiberger, Swaine & Pottenger, 2014, "History of computing"

[86] Freiberger, Swaine & Pottenger, 2014, "Internet"

[87] Borth, 2013, "Mobile telephone"

[88] Borth, 2013, "Mobile telephone"

[89] Poulsen & Boomsma, 2005

[90] Hare & Alloway, 2001

[91] Wild meerkats injected with oxytocin compared to saline controls exhibited a suite of communal cooperative behaviors including digging burrows, colony guarding, pup-feeding and teaching pups foraging as well as exhibiting decreased initiation of aggressive interactions (Madden & Clutton-Brock, 2011). Data showing positive correlations in chimpanzees between levels of prosocial grooming of others (whether or not kinship or sexual relations were involved) and oxytocin levels suggested that these complex social behaviors must be mediated in part by oxytocin (Crockford et al., 2013).

[92] Bethlehem et al., 2014

[93] Stiner et al., 1999

[94] Tomasello & Vaish, 2013

[95] Greene & Haidt, 2002

[96] Hare, Wobber & Wrangham, 2012

[97] Silk, Alberts & Altmann, 2003

[98] Tomasello et al., 2012

[99] Burkart & Van Schaik, 2010

[100] Burkart et al., 2014

[101] Wilson & Holldobler, 2005

[102] Holmes et al., 2007

[103] Diamond, 2002

[104] Testart, 1982

[105] Moll & Tomasello, 2007

[106] McNally, Brown & Jackson, 2012

[107] Geary, 2010

[108] Tomasello et al., 2012

[109] Gavrilets, 2014

[110] Marean, 2015

[111] Wilson & Wrangham, 2003

[112] Tomasello et al., 2012

[113] Dunbar, 1998

[114] Dunbar & Schultz, 2007

[115] Harari, 2014

[116] Murdock, 1945

[117] Piaget, 1932

[118] Strassmann & Queller, 2011

[119] MacLean et al., 2010

[120] Hume 1739

[121] Chang, D'zurilla & Sanna, 2004

[122] This presumes a utilitarian philosophy of ethics that is not universally shared. For Kant, the motivation of the action was what mattered, not its outcome. For Aristotle, the measure of an ethical act was

the extent to which it improved the character of
the person performing it.

[123] Taylor & Moghaddam, 1994

[124] Keegan, 1993

[125] Enquist et al., 2003

[126] Friston, 2010

[127] Ryan & Deci, 2001

[128] Headey, Muffels & Wagner, 2010

[129] Anielski, 2007

[130] Sen, 2009

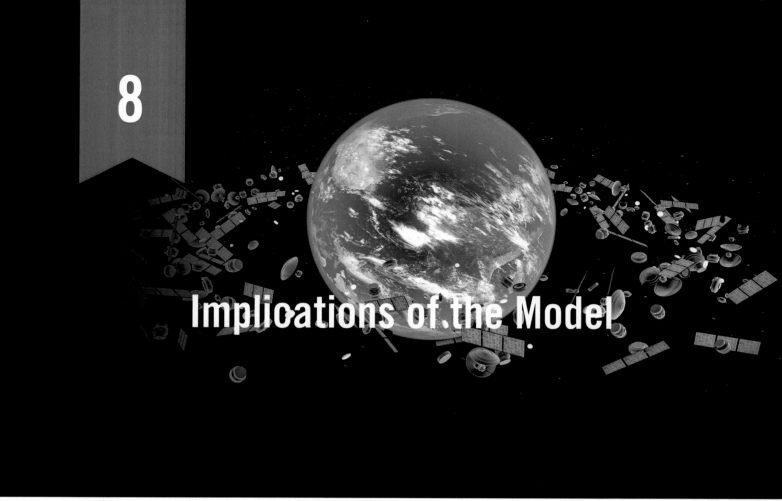

8

Implications of the Model

The energy dissipation model for driving the evolutionary imperative has broad-reaching implications from both the scientific and philosophical perspectives. The following provide some of the more prominent examples of the explanatory power of the model.

8.1 Probabilistic Outcomes, Variation, and Time

Scientists have long debated whether natural processes are directionless (indeterminate) or move toward a particular (determinate) outcome.[1] While our model leads to indeterminate outcomes, it arises from a deterministic set of forces, namely the inexorable drive toward effective energy dissipation, thereby avoiding the fallacy of the exclusive nature of the determinate/indeterminate debate. The model clearly encompasses both predictable and unpredictable elements, resulting in variable approximations to a lawful outcome.

There are several general implications of the energy dissipation model for evolution. The model involves a causal chain operating in unpredictable conditions and modified by recurring feedback rather than a unidirectional cause and effect process. Consequently, the evolving outcome in any particular situation cannot be predicted explicitly.[2] However, its effect on energy transfer can be assessed, and so the probability of potential alternative outcomes can be predicted. This applies equally whether predicting the path taken by a surging flow of lava, the adaptation of a species to a novel challenge, or the pattern of newly learned behavior by an individual organism. Thus, the model provides a framework for accommodating the trajectory of physical processes, the adaptive outcome of natural selection acting on genetic variation, and finally the effects of environmental selective reinforcement acting on behavioral variation.[3]

One of the earliest discussions of the parallel between biological and behavioral selective retention of blind variations was presented by Campbell.[4] The mechanisms of variability and the selection mechanisms are quite different, yet the commonality of their processes across domains is impressive, and so Campbell's thesis stands.

The parallel between changes in a biological character and a behavioral trait is illustrated in Fig. 8.1. A biological character, such as fur thickness in a mammal, shows an approximately normal distribution determined by particular genetic, epigenetic, and developmental constraints that have optimally adapted the animal for the climate typical of its habitat (Fig. 8.1a, blue curve). When that climate shifts toward a colder mean temperature, animals with thinner fur are selected against, while those with thicker fur survive in greater numbers due to natural selection, shifting the distribution of fur thickness for the population upwards (red curve). By the same token, a behavioral trait, such as the use by chimpanzees of a stick as a tool to extract termites from dead logs, may show a normal distribution among individuals within the population (Fig. 8.1b, blue curve). When some chimps fortuitously strip the twig of leaves and bark, and find that the tool thus modified yields more food per unit of effort, the newly-discovered behavior is rewarded (reinforced). Over time, that behavioral trait will be retained with greater frequency through learning within the population (red curve). In both cases, change occurs to optimize the species' ability to survive and consume more energy. Because biological characters and behavioral traits are the product of multiple genetic, epigenetic, developmental, and experiential determinants—the nature of which are subject to ongoing mutation, alteration, or vacillation in an unpredictable manner—the biological or behavioral phenotype that is the product of these determinants is not predictable in any precise fashion. Thus the fundamental unpredictability of the specific nature of the organism's response to a challenge is manifest in the variable output of the genotype and its modulators in species;[5] in the variability evident in even the most repetitive behaviors;[6] and in neural variability generators in the brain.[7] Indeed, such behavioral-neural variability is critical for dealing with unpredictable environmental challenges as seen in the variability of scanning during antipredator vigilance,[8] the organism's exploration of a novel environment, and during classical and operant conditioning.[9]

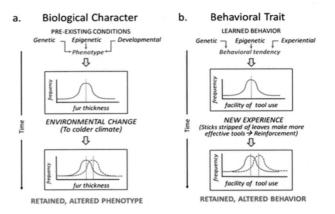

Fig. 8.1. Parallel mechanisms for persisting biological and behavioral changes. (a) The normal distribution for a biological trait shifts when environmental changes increase the selective advantage at one end of the distribution range and decrease it at the other. (b) The normal distribution for a behavioral trait shifts when an animal learns that performance at one end of a variable range of behaviors is advantageous. In both cases, the changes lead to more optimal or complete consumption of energy.

The net variations in behavior seen in these circumstances are not truly random, and so are usually referred to as quasi-random. Congruently, some order has been found in these cases of highly variable patterns of behavior. For example, there are repeated patterns in some variations of the flight paths of fruit flies[10] and consistent individual differences in the degree of variation in the behavior of mosquito fish.[11] Changes in the rate of variation have also been observed in response to environmental contingencies. For example, enhanced rates of genomic variation occur when bacteria are subjected to host immune system attacks on their antigenic substrates.[12] In addition, changes in an animal's rate of emission of spontaneous behavior under stress have been observed.[13] The consequences are an enhanced prospect of evolvability in the former and enhanced rate of learning in the latter. Such biases clearly have adaptive value. Thus, the behavioral variation that is critical for survival and everyday learning is fundamentally parallel to the variation essential for evolution of species by natural selection. And in fact, we do not yet know the extent to which the heritable basis of behavior is subject to epigenetic influences, such as methylation-induced variations in the genes controlling neural plasticity.

The SLT is the only law of physics that refers to the direction of time. All processes are moving from a low entropy state of concentrated energy initiated in the big bang to a state of higher entropy and greater dispersal of energy. The compelling directionality of the entropic process is reflected in the irreversibility of the records of past events. We can all recognize whether a video recording of some event is being played backward or forward. The thermodynamic arrow of time is expressed in the "recognizable" direction of temporal events, e.g., the falling of water off a cliff, the evaporation of vapor from a tree's leaves, and the learning of a child.

Memories of the past and anticipations of the future have been referred to as the psychological arrow of time.[14] This parallel between the entropic and mnemonic worlds has recently been interpreted as due to the absolute dependence of the latter on the former process.[15] That is, the arrow of memory is fundamentally dependent on the thermodynamic arrow of time. It should be added that neurological mnemonic processes of animals have non-neurological equivalents in the genes of plants, the memory banks of computers, and in the stratification and radiation records in rocks in mapping planetary and geological history. The latter provide a critical portion of the evidence for the time course of evolution.

8.2 Gaia and Medea as Global Models of Change

The global Gaia process is a model that proposes a homeostatic equilibrium for evolution in which the abiotic and biotic worlds constitute an integrated biosphere optimally suited for perpetuating life on Earth.[16] Evidence for this process lies in the buffering capacity of the biosphere, the complexity of the web of life, its sensitivity to unpredictable variations of the environment, and the exquisitely narrow range of the universal constants supporting the structure of the universe and life itself. According to this model of the biosphere, the entire planet acts as a living organism, to process energy in a steady-state manner that maintains its viability over time.

By contrast, the Earth's history as a Medean succession of cataclysmic extinctions of life clearly identifies exceptions to the view of Earth as a benign and nurturing "Mother."[17] Instead, the planet is seen as host to an ecosystem that is fragile and susceptible to radical change, through partial to near total extinction of its various forms of life, in response to chance, catastrophic events. In fact, both sets of observations (Gaian and Medean) are congruent with the thermodynamic model proposed here. Whether driven by the steady pace of gradual evolution in a homeostatic biosphere, or destabilized by periodic near-extinctions in the wake of catastrophic events, the net effect is progression toward a world that leads to ever more effective energy flows.[18] As noted throughout this book, the thermodynamic interpretation of change includes both the biotic and abiotic worlds. This model embraces both. In a word, energy gradients are causes and energy flows are effects. The biosphere in its totality, whether gradually or precipitously, changes through time in response to the thermodynamic imperatives of the SLT and the PLA.

8.3 Emergent Phenomena

The hierarchy of the levels of scientific focus runs from particle physics, through many body physics, chemistry, molecular, organismic, and population biology, to the behavioral and social sciences. The model's built-in variability and probabilistic predictions permit the emergence of new properties at any scientific level. Because of this and because the premises of the model can be directly applied at any level of the system, the challenge of explaining or predicting emergent phenomena through levels of the scientific hierarchy from a reductionist versus a holistic perspective is avoided.[19] As Pernu and Annila argue, if natural events consist of energy flows between each entity and its environment, then emergence can be seen as a natural outcome guided by the SLT and PLA regime of effective consumption of free energy. Thus emergence is an expression of the cumulative effects of myriad natural interactions.

When a new source of energy becomes available, the PLA predicts that new biological forms will evolve to exploit it. The change may be so enduring and remarkable that it represents a novel form—an emergent event. Yet its particular manifestations cannot be predicted in advance. The appearance of eukaryotic cells two billion years ago has been related to rising oxygen levels.[21] The greater efficiency of aerobic metabolism permitted the evolution of organelle formation, cell compartmentalization, nuclear signaling systems within and between cells, the consignment of respiration to mitochondria, and the emergence of multiple cell types.[22] The particular form that eukaryotic cells would take, however, was not predictable merely from the abundance of oxygen.

The degree to which human evolution has been shaped by energy transformation in the form of information creation and exchange is a dramatic example of an emergent phenomenon. The social intelligence view of evolution argues that rather than simply evolving a greater degree of general intelligence, humans have evolved specialized social intelligence (as detailed in Section 7.9). Humans have used information processing for the evolution of cultural intelligence well beyond that of apes.[23] Humans have developed specialized social-cognitive skills for living and for exchanging knowledge in cultural groups: communicating with others, learning from others, and reading the mind of others in especially complex ways. Thus emergent cognitive propensities underpin the emergence of information exchange as the singular human form of energy transformation. See Section 8.5 for elaboration of these cognitive mechanisms.

In the strictly materialist tradition, the mind is viewed as an emergent phenomenon of neural activity. We suggest that the supreme human capacity for dissipating energy in the form of information transfer offers a perspective, if not a resolution, of the historical question posed by the mind-body problem. The postulation of mind-body dualism in its traditional form rests on the concept that mind has different properties from the body and that those properties are not interchangeable.[24] The properties of the mind have been distinguished from those of the body by the mind's nonphysical nature, its indeterminacy, and its subjective aspects, such as qualia or raw feels. Critically, the current model posits that both mental activity and behavioral actions are indeed fundamentally involved in energy dissipation, are unpredictable, and are able to assess sensory events like other bodily actions and the neural processes that subserve them. Thus the model provides a unifying perspective encompassing both the mental and physical worlds by offering an alternative to the historic conundrum of their incommensurability, i.e., the mind-body dualism of Descartes and Kant.[25] The application of quantum modelling to both cognitive and neural processes supports this perspective. Similarly, this perspective is compatible with the trend in modern psychiatry to recognize that the stressful experiences of the patient, as reflected in the patient's behavioral changes, are manifestations of chemical changes in the patient's nervous and endocrine systems and in the epigenetic changes in the patient's genome. The historical gulf between mind and brain loses its significance as changes in each system reflect and feed back upon the changes in energy dissipation in the others.

8.4 Increased Complexity, Diversity, and Hierarchy as Evolutionary Directions

Another example of emergent properties are evolutionary changes that tend toward increased complexity, diversity, productivity, information content, hierarchical organization, and homeostatic control. These emergent properties occurring over eons are so impressive that they have been perceived as candidates for the imperative of evolution itself. Our view is that persisting trends are not directional driving forces for evolution. Rather they are manifestations of the SLT and the PLA generating ever greater energy dissipation.

There have been numerous proposals that evolution is devoted to building complexity, diversity, or hierarchical relations.[27] Chaisson[28] has provided compelling evidence for the *net* growth of complexity in, first, the physical world, then in the living world over time. We believe that this net increase in the rate density of energy flow, whether by a galaxy, organism, or group of organisms, is a direct reflection of the SLT and the PLA in action. Averaged over eons across galaxies, solar systems, species, or cultures, we believe that changes in the aggregate reflect increased complexity because the natural tendency for energy flow to increase is what drives evolution in that direction. A common expression of that increased flow is often, though not always, increased complexity.

Gould[29] listed three questions as central to attempts to describe the history of life: (1) What is the direction of life's history? (2) Is organic change directed by the environment or by organisms? (3) Is the change gradual or episodic? On the second question, Gould argued for an interactive version involving the contribution of variation from both the environment and from organisms; and on the third question, he espoused the episodic position (punctuated equilibrium). In this book our central focus is on the first question. Regarding the direction of evolution, Gould[30] doubted the likelihood that complexity or diversity were adequate descriptions of the direction of life history because, although there is impressive evidence supporting these trends, there are also many examples of reversals of these trends, with complex features being simplified and species being extinguished. To accommodate these disparate observations, Gould concluded that the trajectory of evolution is the consequence of random variation.[31]

The *Zero-Force Evolutionary Law* (ZFEL) of McShea and Brandon[32] poses a similar premise with a different emphasis. The ZFEL states that "In any evolutionary system in which there is variation and heredity, there is a tendency for diversity and complexity to increase." That tendency "is always present but may be opposed or augmented by natural selection, other forces, or constraints acting on diversity or complexity." In other words, if variation is a property of the system, the tendency over time will be for that variation to generate diversity through a random process. In their view, change is the default (zero-force) condition of the natural world, with increasing diversity its inevitable consequence. Complexity tends to increase as well—by definition, since complexity is defined simply as the total number of differentiated parts. Assuming the changes that generate complexity and diversity are heritable, they accumulate over time. While natural selection and other forces or constraints may alter the nature of diversity and complexity, nothing other than random events are necessary to explain both diversity and complexity, in this view.

Neubauer rejected Gould's view[33] that randomness is the only underlying cause of the observable increase in diversity and complexity, as do we. While we tend to agree with the logic of McShea and Brandon's argument that the ZFEL can explain the growth of diversity over time, we don't agree that it adequately explains the increase in complexity, since their definition of complexity, in our view, is too minimalistic. Neubauer defined complexity more realistically as *the number of parts and the number of functional relationships among those parts*.[34] Using energy rate density (defined in Section 4.4) as a proxy for complexity, Neubauer believed that increases in diversity and complexity are driven by the bias of the SLT toward forms that support ever more energy dissipation. In reviewing the literature on evolutionary changes over the eons with his definition of complexity in mind, Neubauer concluded that in general, evolution has produced an ever greater diversity of elements (types of cells, organs, genetic families and species) and greater complexity of organization (brain organization, gene interactions, information exchanges, life styles, varieties of predation, the organization of the food web, and cultural organization).[35] He did not refer to the PLA, however, which is critical in adding an optimization imperative to the increases in energy flow prescribed by the SLT and the PLA.[36]

Linking his conclusions with the data of Chaisson (cited previously), Neubauer provided an impressive account of how the complexity and diversity of species arose and how these emergent properties serve the mandate of the Second Law. Appropriately, Neubauer cautioned that not all of nature is striving toward greater complexity. What nature is achieving is the optimal flow of energy through the ecosystem, consistent with the survival of its components, constrained by the happenstance of historical trajectories, and responsive to the vagaries of random (or at least unpredictable) events. We think the overwhelming evidence for the PLA is critical in formally legitimizing this conjecture.

Were the direction of evolutionary adaptation to be *inevitably* toward complexity or diversity (and thus toward hierarchy, in some cases), exceptions to the growth in complexity over time should be lacking. Yet they are not. Some environments have undergone minimal increases in the number and diversification of species, showing little if any increase in the variation of forms over millions of years, as at deep ocean depths.[37] Furthermore some living forms have become simpler rather than more complex over millions of years, showing regressive evolution, or emergent simplicity rather than complexity. Examples include evolved brain simplification in salamanders;[38] loss of eyesight in many cave dwelling forms;[39] evolutionary simplification of shells and other exoskeletons across many metazoa;[40] consolidation of skull bones across vertebrates;[41] reduction in the pelvic girdle of whales and dolphins;[42] reduction of the wings of flightless birds;[43] the loss of sweet receptors and sweet preferences in exclusive meat eaters, such as sea lions, spotted hyenas, and bottlenose dolphins;[44] and the loss of complex behavioral traits in crabs, mice, and carnivores.[45] Many of these cases involve simplification of energetically expensive developmental programs for anatomical features no longer needed. In others, they represent the adoption of alternative features that enable more efficient energy acquisition or consumption. In the case of the loss of eyesight in Mexican cave fish, for example, the regression has been demonstrated to reflect natural selection for overall metabolic savings.[46]

Parenthetically, scholars who espouse special creation often cite the mammalian eye as an example of such irreducible refinement that it could not possibly be the result of evolution guided by the quasi-random forces of natural selection.[47] That special creation would allow regression of this wonderfully wrought organ to a vestigial nubbin is as difficult to understand as it is to comprehend how the blinding mentioned above is an example of increasing complexity. However, congruent with the current perspective, these evolved simpler forms are measurably more effective at energy dispersion in their particular environments, requiring fewer resources to process more energy, than their more complex ancestors or than other species in their environment.[48]

The energy dissipation model advanced here predicts that extremely complex forms incorporating hierarchical integration, as has happened in humans, will evolve if they can access energy sources unavailable to simpler forms.[49] However, this elaboration is not unrestrained, because increased complexity carries with it an increased likelihood of malfunction.[50] Typically, a more moderate degree of complexity emerges as the most adaptive.[51] Chaisson explained this by emphasizing adaptive *optimization* of energy flow, whereby energy dissipation is most advantageous when it proceeds at an optimum rate between that which is too small to effect change, and a rate so great as to be destructive. Excessive rates of resource depletion will result in the disintegration, remodeling and perhaps eventual extinction of complex forms, including the collapse of human cultures.[52]

In summary, diversity, complexity, and hierarchy are most parsimoniously viewed as circumstantial emergent processes whose involvement and expression vary with the setting and the phenomenon under inspection. They are common trajectories, but are collateral effects of the fundamental, overriding drive of the SLT and the PLA toward optimal dissipation of energy.

8.5 Heuristic versus Analytic Cognition

In cognitive studies, a useful if simplistic distinction has been made between heuristic and analytic modes of thinking. The two cognitive states differ not only in their behavioral attributes but in their consumption of energy.

Heuristic thinking relies on non-conscious processes including feelings and emotions.[53] It is the default mode, is used much more frequently, is intuitive, is relatively effortless, is more likely to be selfish and prejudiced rather than prosocial, depends on well learned rules or on cognitive biases, and involves minimal exercise of will power.[54]

Analytic thinking involves the will power that comes from exercising self-control in the form of conscious effortful self-regulation, as in focusing attention, analytic decision making, developing, anticipating and achieving a planned outcome, overcoming selfish impulses and old habits, and engaging in social negotiation and compromise.[55]

In contrast to heuristic thinking, analytical thinking is relatively more time consuming and energetically costly.[56] Energy consumption by brain dopamine circuits is involved in cognitive planning,[57] and stimulation of these circuits has been shown to increase brain glucose utilization in the rat.[58]

In most situations, heuristic thinking provides a workable approximation to an effective action plan because the rules guiding the behavior are amenable to automatic updating by past success and failures. Since food is susceptible to periodic or chronic limitation for most species in nature, the workable control of behavior by the lower energy demands of heuristic cognition favor it as the default cognitive strategy. Those species in which analytic cognition has assumed a larger role typically benefit from the availability of an ample or higher-energy diet.

8.6 Volitional Control and Free Will

The ability to vary one's own behavior or behavioral degrees of freedom, contributes to the subjective experience in humans of volitional control or free will.[59] This is not to suggest that free will is illusory or without survival value. The exercise of free will is energetically expensive (see below) and so it is highly unlikely that such an exception to the Second Law would arise without having some selective advantage. So the alternative hypothesis is that free will (to the extent that it does exist) is adaptive and consonant with the Second Law. This is congruent with the common experience that the exercise of free will, far from being a fictional illusion is universally perceived as effortful and as commanding our limited attentional resources. It is significant that animals show active intentional behavior in arriving at decisions involving multiple options.[60]

Although we have no idea what the subjective experience of volition is like in other animals, the planning and carrying out of a decision based on weighing of alternatives appears to be an evolutionary adaptation. If so, the human variant is presumably more complex than that of other species. Indeed the evolutionary development of free will in humans has been hypothesized to afford individuals the advantages of joining in complex social interactions in pursuit of increased energy options and resources.[61] If so, the proposed evolutionary

advantage of collective action in securing food should also be observable in more rudimentary forms in animals. This is the case in both prey and predatory species.[62] Indeed, experimental studies in humans have found evidence that the exercise of free will affords several forms of behavioral control critical for guiding an individual's social behavior, i.e., self-regulation, rational choice, ad hoc planning and taking initiatives.[63] The exercise by individuals of these manifestations of free will are critical for regulating thoughts, appetites and impulses, accepting societal roles, and performing social duties, including sacrificing oneself for the group. While the survival advantages for members of groups are obvious, it is also true that individuals with high self-control succeed better than those with low self-control in school, work careers, and in personal adjustment.[64] In fact, self-control trumps IQ and self-esteem in predicting school grades.[65] As individuals differ with regard to self-control, it isn't surprising that personality differences are related to basal metabolic rates in many species.[66] In the context of the present model, the evidence then suggests that free will is highly evolved in humans to ensure compliance with the mandate of the SLT and the PLA in effecting maximal energy dissipation.

A subjective sense of loss of volitional control often accompanies various forms of illness.[67] Such impairment may be expressed as a generalized fear and resistance to change or more seriously as mental pathology, and is manifest in the iterative thoughts and recursive behaviors of many anxious, angry, depressed, post-traumatically stressed, schizophrenic, and obsessive compulsive individuals.[68] Their ideation fits a pattern of reduced intake of energy or information (self-absorbed, isolationist, avoidant, and immersed in their own emotion and thoughts) and reduced energy expenditure and/or information utilization (increased heuristic, iterative thought, acting impulsively on biases, recursive behavior, minimally coping with stressful events or environmental change, and reduced planned activity, productivity, or consumption).[69] These conditions of low energy flow are experienced as unhealthy and unpleasant (anxiety-laden, lonely, depressing, confusing, and irritating) and are characterized by reduced cerebral blood glucose consumption and reduced brain activity.[70] In the brain, receding conscious awareness has been shown to follow decreased brain glucose consumption with accompanying decreased glutamate cycling and reduced differential activation of brain regions mediated by waves of dendritic synchrony.[71] Reduced insulin sensitivity has been shown to be a mediator between body mass index and reduced activation of the right parietal cortex, impaired memory in obese adults,[72] and in Alzheimer's disease.[73]

Treatment for mental illness may be seen as restoring the individual to a normal level of energy input and output, increasing behavioral variability and self-control.[74] Exceptional mental pathologies include those that increase energy intake and output for personal gain beyond the socially accepted in-group norms, as in addiction, manic behavior and psychopathy. Congruently, these states are experienced as not unpleasant and show elevated levels of blood glucose in brain reward structures, such as dopamine mediated effects in the nucleus accumbens.[75] Chronic activation of the reward pathways (primarily dopaminergic and opioid) as seen by indulgence in high caloric foods or recreational drugs, leads to the enhanced response of these structures to addictive cues, chronic craving and overconsumption.[76] It is not surprising that a maternal diet of energy rich foods activates these circuits and induces preferences for such foods in the young.[77] The mandate of the Second Law to dissipate energy is conveyed powerfully by the evolved reward systems of the brain. The ramifications for pathologies of over indulgence are apparent, as in the case of obesity and addiction.[78]

If the exercise of free will is energetically expensive, it is likely to be engaged when energy is readily accessible for uptake. This is the case when resources are plentiful. Thus in the classical meaning of free will as a state in which the individual is able to fully express options in exercising degrees of freedom of action,[79] increased degrees of freedom of choice are afforded by increased availability of free energy. Conversely, when circumstances are demanding, and access to free energy is limited, freedom to act is limited, and heuristic rule-guided thinking predominates. Congruently, if energy reserves are depleted by engaging in analytic thinking,

the individual is more likely to default to heuristic thinking, such as behaving aggressively in situations calling for social restraint.[80]

8.7 The Meaning of Life

Life has meaning if it has a purpose. The purpose of life has been the subject of philosophical and religious contemplation since ancient times. "Purpose" may be viewed as a convenient, abbreviated way of referring to causal sequences which have occurred in the past and have predictive value. So in what sense can any purpose be attributed to life?

The one thing that all forms of life *do* is dissipate energy. Just as one can say that the chicken is the egg's way of making another egg, one could argue that the logical implication of our thesis is that life is energy's way of degrading itself. But does that make energy degradation the *meaning* of life?

Applying this to the present issue, scientifically speaking, we would argue that dissipation of energy is not the purpose, but rather is the ultimate cause of the evolution of life and the force driving change in the natural world. The difference is apparent when we think of a stream flowing down a hill or a bird building a nest as being manifestations of the ultimate cause—energy dissipation—as opposed to the stream or bird having an intentional purpose, or for that matter, free will. The term "explanation" of life is more congruent with the causal facts. Expenditure (dissipation) of energy gives life meaning, in the sense that it explains life's existence.

We realize that this philosophical argument leaves some people unconvinced. Many undoubtedly view the scientific account of evolution as spiritually unsatisfying, counterintuitive intellectually, and dehumanizing in robbing them of any sense of direction or control in their personal lives. However, the present model actually provides compelling evidence that evolution has direction and does produce orderly predictable consequences. Evolution brings about changes in the animate and inanimate realms that result in an increase in the transformation of energy. Humans are at the apex of this trend for more recently evolved life forms to transform ever more energy, and to that extent do have some control over the way that evolution plays out. Indeed the model recognizes the exceptional nature of the human contribution to shaping the world. From this perspective, the ultimate or fundamental meaning (explanation, or purpose) of life lies in the understanding that life can be viewed as a system for effectively transforming energy.

8.8 The Religious Perspective

A critique of our model by some will likely be that it fails to address the feelings that humans have about their lives as they experience them. It does not provide a platform for their feelings of gratitude over the positive aspects of their lives, nor provide comfort for their afflictions. Neither does it accommodate the wonder they feel about the beauty and mysteries of the world around them, nor help them anticipate the future in any large sense. These are roles that religion plays, perhaps contributing as much as in-group cohesion to the pervasiveness of religion in all the cultures of the world.

While this treatise is intended as a strictly scientific explanation for the factors and forces that impel evolutionary trajectories in the physical and living worlds, it is not our intent to ignore the perspective provided by religious revelations and teachings on the subject. The following is a brief review of how different religious traditions have embodied some of the concepts that we have heretofore examined from a purely scientific point of view.

We have provided evidence that all living things have higher levels of energy consumption and storage than the air, water and land around them. Even seeds, spores, and ectothermic animals show greater energy turnover than their surroundings. While it is alive, an organism takes up and stores energy from the environment, and of course, expends energy in the process of living. If this is so, associating energy with the divine is not an illogical assumption.[82]

In a similar vein, the increasingly important role that information is playing in the ways that human society is evolving finds resonance in the beliefs of the ancient Greeks, who held that there were two forms of knowledge, *mythos* and *logos*. To the ancient

Greeks, the term "logos" meant word, speech, or reason. Philo of Alexandria (20BCE-50CE), who sought to reconcile the Greek and Judaic traditions, believed that "logos" was synonymous with God and that "logos" in humans was reflected in the deliberations of rational thought.[83] If we follow Philo in believing that the essence of God is rational thought, its current equivalent is analytical thinking. As noted above in Section 8.5, analytic thinking is the most energy dissipative form of mental activity. The other connotation of logos is, more broadly, information. As we have noted throughout the book, information dissipation is the key to understanding the pre-emptive role of humans in energy dissipation.

While our treatment of this subject may leave some readers disquieted because it does not address their personal emotional issues regarding morality or their moral life choices and experiences, it may be helpful to go with the ancient Greeks again in distinguishing the two forms of knowing, mythos and logos. The Greeks regarded both mythos and logos as sources of truth. But mythos was conveyed in stories, poetry, lyrics and music. While these also were sources of truth, they further served as sources of emotional perception, insight, heuristical understanding, theology, therapy, social bonding, and entertainment. We discussed the functional significance for life and evolution of this mode of expression in the previous chapter.

Finally, we would point to the congruence of our model with that of the current perception of two religions on the implications of the human rush to expend ever more energy, with its impact on climate and the ecology of the planet. We introduce it here while on the subject of whether our model speaks to moral issues as effectively as religions. We turn to recent statements by two religious leaders on the moral issue of climate change.

Buddhism's 14th Dalai Lama has expressed concerns about the consequences of our profligate behavior for the environment.[84]

Human activity everywhere is hastening to destroy key elements of the natural ecosystems all living beings depend on. Ignorance of this interdependence has harmed not only the natural environment but human society as well. We have misplaced much of our energy in self-centered consumption, neglecting to foster the most basic human needs of love, kindness and cooperation. This is very sad. Our Mother Earth is now teaching us a critical evolutionary lesson—a lesson in universal responsibility. Morally as beings of a higher intelligence we must care for this world. Its other inhabitants—members of the animal and plant kingdoms—do not have the means to save or protect it. Let us adopt a lifestyle that emphasizes contentment, because the cost to the planet and humanity of ever-increasing "standards of living" is simply too great."

In a similar vein we see in Sections 67 & 68 of Pope Francis' recent encyclical "Laudato Si" a description of past human behavior and its likely consequences. His perception is congruent with that of our model.

> 68. This responsibility for God's earth means that human beings, endowed with intelligence, must respect the laws of nature and the delicate equilibria existing between the creatures of this world, for "he commanded and they were created; and he established them for ever and ever; he fixed their bounds and he set a law which cannot pass away" (*Psalms* 148:5-6, ibid). The laws found in the Bible dwell on relationships, not only among individuals but also with other living beings. "You shall not see your brother's donkey or his ox fallen down by the way and withhold your help… If you chance to come upon a bird's nest in any tree or on the ground, with young ones or eggs and the mother sitting upon the young or upon the eggs; you shall not take the mother with the young" (*Deuteronomy* 22:4, 6, ibid). Along these same lines, rest on the seventh day is meant not only for human beings, but also so "that your ox and your donkey may have rest" (*Exodus* 23:12, ibid). Clearly, the Bible has no place for a tyrannical anthropocentrism unconcerned for other creatures.

Chapter 9 of this book is in complete agreement with the assessments of the Dalai Lama and Pope Francis on the effects of our profligacy. In fact the model goes beyond the solutions they propose to suggest that understanding of the nature of the process offers a solution with some chance of being successful in addressing the problem. This is not to deny that religions have an essential role to play. The common values of many religions for reverence, respect, restraint, redistribution, and responsibility can provide critical momentum.[86]

8.9 Life on Other Worlds

Evidence against the operation of the SLT and PLA has never been found. Indeed, there is plenty of evidence for the influence of the SLT and PLA in the inanimate non-terrestrial world of galactic and planetary space. As indicated in Chapters 3 and 4, examples include the evolution of galaxies, galactic rotation, the evolution and death of stars, planetary system formation, the orbital behavior of planets, planetary plate tectonics, planetary drainage patterns, weathering and planetary weather vortices. At the micro level, examples include the evolution of chemical elements, chemical minerals, and thermosynthetic reactions. With the apparent ubiquitous influence of the SLT and PLA driving these systems to ever more optimal dissipation of energy, it is reasonable to assume that the ultimate expression of these forces—namely the appearance of life on earth—should also be evident on worlds beyond Earth.

The assumption that life exists on other worlds is not new. Early Greek philosophers, such as Democritus, held that the Milky Way galaxy consists of multitudes of stars with their own planetary systems sustaining life.[87] In 1584, Giordano Bruno wrote that there must be ". . . countless suns, and countless earths all rotating around their suns."[88] However, certain knowledge of planets orbiting other stars has been acquired only within the last three decades. After discovery in 1992 of the first of these planets[89]—called "exoplanets" since they exist outside our solar system—a cascade of others has followed, so that more than four thousand exoplanets had been discovered by the end of 2020.[90] It appears now that exoplanets are so common that some astronomers believe nearly every star has one or more exoplanets in orbit around them. Since the methodology for discovering exoplanets inherently conveys certain information about their characteristics—like orbital period, mass, and distance from their central star—a scientific assessment of the possibility that life could exist on them has become theoretically possible for the first time in the multi-millennial history of speculation about life on other worlds. And such an assessment is overwhelmed from the start by the sheer number of possibilities.

Assuming that the conditions which gave rise to life and spurred its evolution on Earth operate in the same way throughout the universe, and conceding that exoplanets abound in profusion, even if they don't orbit every star, the possible number of sites where life could exist beyond our home planet is close to unimaginable—so much so that even if a tiny fraction of exoplanets does actually harbor life, the total number that does so has to be staggering. Modern-day astrobiology (study of the origin, evolution, prevalence, and future of life in the universe) is in roughly the same position with regard to alien life that the educated world was with regard to the shape of the globe in the year 1520—every naturalist and knowledgeable mariner knew the world was round, even if no one had yet to circumvent it.

Is it possible to get beyond mere amazement to some actual predictions about the frequency and nature of life on other worlds? The simple answer is "yes," with the huge caveat that the degree of uncertainty in all such estimates is very high.

One of us (Irwin) and his colleagues[91] have analyzed a database of confirmed exoplanets for which thermal models are available, and calculated a "biological complexity index (BCI)" that calculates the relative probability that conditions conducive for the evolution of life beyond the microbial level are present. They concluded that 1.6% of exoplanets may be capable of supporting life beyond the microbial level of complexity. Based on more conservative criteria, Buonama and her colleagues[92] estimated that the number of planets on which complex forms might have evolved would likely be about a thousand-fold smaller (0.002%).

Taking the extremes of these estimates, based on liberal (1%) or conservative (0.002%) constraints respectively, the number of such planets in our galaxy that could be hosting complex forms of life can be calculated at somewhere between 2 million and 1 billion.[93] There is no way of estimating what fraction of those biospheres has seen the emergence of technologically adept beings capable of interstellar communication and travel. If 1 in 1000 of them has done so, there could be between 2,000 and 1 million planets in our galaxy capable of travelling or communicating across interstellar space.

The central thesis of this book is antithetical to the view that we, alone among the millions if not billions of planets capable of sustaining life in the galaxy, or the billions if not trillions of planets capable of doing so throughout the cosmos, are the only instance of technologically advanced evolution in the universe. Since technology is the most effective and efficient way to consume energy and increase entropy yet to evolve on our planet (Section 7.1), there is every reason to assume that the PLA has favored its evolution on other worlds where circumstances and resources make it possible. And the number of such worlds, whatever the constraining assumptions, are now sure to be overwhelmingly large, if very far away.[94]

Summary of Chapter 8

The energy dissipation model for evolution propounded here has far-reaching implications at many levels, and strong explanatory power for many phenomena.

Both the physical and living worlds appear to operate as complexly interacting systems, driven by a causal chain that operates under unpredictable conditions and is modified by recurring feedback, rather than by a unidirectional cause and effect process. The complexity of these interactions is superimposed on the inherent probabilistic properties of nature, ranging from quantum effects at the subatomic level to the variability of behavioral and social interactions. Consequently, the way in which any particular situation unfolds cannot be predicted explicitly, even though the overall consequence of change is the predictable degradation of energy in the most effective way possible.

Viewed from a global perspective, the planet as a whole persists as an undying, stable, biosphere that continually consumes and degrades energy. Whether viewed as an ecosystem in harmony with its environment, floating through space as a homeostatic, self-sustaining mega-organism (the Gaia model), or as a home for tenuous life that periodically is driven to near-extinction by cataclysmic events (the Medea model), the planet exists as an island in the vacuum of space, where energy is focused, consumed, and degraded to drive everything that is dynamic about our world.

The details of change cannot be anticipated precisely, so new forms and functions arise that are not predictable from their antecedents. The insertion of natural selection into the process of organic evolution, coupled with inherent variation, random mutation, and the transmission of cumulative genetic information across generations, ensures that new organisms with novel mechanisms for the transformation of energy will emerge over time. The novelty of emergent phenomena, spurred by the imperative to transform more energy more effectively, reflects the fact that evolution at all levels is a creative process.

Evolution's tendency to generate diversity over time can be viewed as the general rule rather than the exception (though exceptions do exist). In a dynamic world, where change arises from preexisting forms and is preserved by genetic retention along with its predecessors, diversity inevitably ensues. This has been referred to as the *zero-force evolutionary law*, which may be substantially sufficient to explain the evolution of diversity. Why evolution has led to greater *complexity* overall within the living world, however, requires a more complicated rationale. Our view boils down to the fact that increased complexity, on average, leads to greater and more effective energy turnover, which usually (though not always) provides a survival advantage that is favored by natural selection. The exceptions, such as evolutionary simplifications in static environments and structural consolidations that improve survivability, help to prove the rule.

Animal behavior can be viewed as controlled by two categories of cognition, here distinguished as heuristic or analytic thinking. Heuristic thinking relies on non-conscious processes and is weighted by feelings and emotions. It is used much more frequently, is intuitive, is relatively effortless, is less demanding of brain glucose, is more likely to be selfish and prejudiced rather than prosocial, and depends on well learned rules and cognitive biases. Analytic thinking, by contrast, consists of conscious effortful self-regulation, as in focusing attention, analytic decision making, developing, anticipating, and achieving a planned outcome, overcoming selfish impulses and old habits, and engaging in social negotiation and compromise. When circumstances are demanding and access to free energy is limited, heuristic rule-guided thinking predominates. Since this is the condition most commonly confronting animals in nature, heuristic cognition is the default mode of behavioral control. Species in position to take advantage of the superior survival value of analytic thinking typically have access to greater energy resources. Congruently, if energy reserves are depleted by engaging in analytic thinking, the individual is more likely to default to heuristic thinking, such as behaving aggressively in situations calling for social restraint.

An evolutionary imperative driven by the SLT and channeled by the PLA sheds light on the ancient philosophical conundrum of determinism versus free will. The fact that evolutionary trajectories are driven by these thermodynamic mandates underlies the ultimate deterministic nature of all changes. Were every factor that influences behavior known— every input into the nervous system of an animal and every neural mechanism that generates a behavioral output precisely definable—then behavior itself could be argued to be strictly deterministic. But not every input is known, nor are the mechanisms by which the inputs are processed and linked to emitted behaviors fully understood. This fact, superimposed on highly variable circumstances, and different historical antecedents, plus the vagaries of individual variation, combine to give the appearance that much of behavior is freely determined and volitional, or goal directed and self-initiated, at least in humans and in many animals. Indeed, the ability to behave in a volitional way and to choose among alternatives is clearly an adaptive advantage. Regardless of the extent to which behavior is ultimately deterministic, it operates as if alternatives are freely chosen with specific purposes in mind. The organism doesn't have to be mindful of the ultimate thermodynamic imperatives of its actions, for those imperatives to be a fact.

Another philosophical implication of the energy degradation model of evolution derives from the fact that all change is driven by energy transformations. This doesn't make energy transformation the *meaning* of life, any more than resolving potential energy gradients due to gravity is the *meaning* of a stream that flows downhill. The significance of the stream, in terms of its practical value or esthetic virtues, can be appreciated with no reference whatsoever to gravity; but the force of gravity is the ultimate explanation of why the water flows downstream. Likewise, the evolution of life has given rise to a proliferation of "endless forms most beautiful and most wonderful" in the words of Darwin,[95] which need no reference to the vital role that energy plays in the process of being alive. The human experiences of gratitude, wonder, love, hate, fear, hope, and anticipation—the traditional province of religion—do not derive in any obvious way from thermodynamics. Yet the current views of two of the worlds' religious leaders are quite congruent with the conclusions of the present model regarding the importance of dealing with the effects of our increasing expenditure of energy on the environment and climate change. The essence of life ultimately does boil down to the flow of energy through an incredibly complex array of molecules. So while the *meaning* of life can be appreciated differently from a multitude of viewpoints without reference to energy, everything that living organisms do is animated ultimately by its consumption; and all the anatomical elegance, physiological efficiency, adaptive ingenuity, and purposeful behavior of living organism reflects the operation of the SLT and the PLA through the evolutionary process. If the transformation of energy is not the meaning of life, it provides an explanation and forecast which leads to a testable operational set of recommendations.

Finally, while all the considerations above are based on observable facts pertaining to planet Earth, there is no reason to doubt that the same principles apply throughout the universe. Accordingly, our inferences about the nature of evolution, including the inevitability of life's emergence, surely pertain to other worlds. The extremely large number of planets now known to exist means that life exists on some of them with near certainty, and in some cases is likely to have evolved to a level of complexity and technological capability equal to or beyond our own.

References and Notes

[1] Monod, 1971; Gould, 1977

[2] Annila & Annila, 2008; Annila & Kuismanen, 2009

[3] Beck, Cooper & Carter, 1994; Sharma & Annila, 2007

[4] Campbell, 1960

[5] Feinberg & Irizarry, 2010; Perry et al., 2007

[6] Devenport, 1983; Page & Neuringer, 1985

[7] Devenport, Hale & Stidham, 1988; Brembs et al., 2002

[8] Bednekoff & Lima, 2002

[9] Devenport, Hale & Stidham, 1988; Shahan & Chase, 2002; Brembs, 2011; Devenport, 1983

[10] Maye et al., 2007

[11] Biro & Adriaenssens, 2013

[12] Graves et al., 2013

[13] Devenport, 1983

[14] Hawking, 1985

[15] Mlodinow & Brun, 2013

[16] Margulis & Lovelock, 1974; Lovelock & Margulis, 1997

[17] Raup & Sepkoski, 1982; Ward, 2009

[18] Karnani & Annila, 2009

[19] Anderson, 1972

[20] Pernu & Annila, 2012

[21] Dismukes et al., 2001

[22] Baudouin-Cornu & Thomas, 2007; Stamati, Mudera & Cheema, 2011

[23] Wright, 2000; Herrmann-Pillatha & Salthe, 2011 Ridley, 2010

[24] Fodor, 2006

[25] We do not suggest, however, that the "hard problem" of philosophy (how to explain subjective experience) is thereby illuminated.

[26] McShea, 1991; Goodwin, 1994; Morowitz, 2002; Chaisson, 2009; Mitchell, 2009

[27] Salthe, 1985; Kleidon, 2010

[28] Chaisson, 2001

[29] Gould, 1977

[30] Gould, 2000

[31] Gould, 1996

[32] McShea & Brandon, 2010

[33] Gould, 1996

[34] Neubauer, 2012

[35] Neubauer, 2012

[36] Karnani & Annila, 2009

[37] Saunders, 1979

[38] Roth, Nishikawa & Wake, 1997

[39] Culver & Sket, 2000

[40] Aleshin & Petrov, 2002

[41] Sidor, 2001

[42] Fordyce & Barnes, 1994

[43] Elliott et al., 2013

[44] Jiang et al., 2012

[45] Whishaw et al., 2001; Iwaniuk & Whishaw, 2000

[46] Protas et al., 2007

[47] Forrest & Gross, 2004

[48] McShea, 1991; Silva, Latorre & Moya, 2001

[49] Salthe, 2004; Friston, 2010; Annila & Salthe, 2010

[50] Salthe, 2010

[51] Wang, Liao & Zhang, 2010

[52] Diamond, 2005

[53] Kahneman, 2003; Baumeister et al., 2008

[54] Simonson, 2005; Masicampo & Baumeister, 2008; Gailliot et al., 2009; Pocheptsova et al., 2009

[55] Baumeister et al., 2008

[56] Gailliot et al., 2007; Baumeister et al., 2008

[57] Schott et al., 2008

[58] Esposito et al., 1984

[59] Brembs, 2011; Frith et al., 1991

[60] Seth, Baars & Edelman, 2005; Trewavas, 2009

[61] Baumeister & Masicampo, 2010

[62] Pitcher, Magurran & Winfield, 1982; Partridge, Johansson & Kalish, 1983

[63] Baumeister, Vohs & Tice, 2007b; Dewall et al., 2007; Baumeister, Crescioni & Alquist, 2010

[64] Mischel, Shoda & Peake, 1988; Tangney, Baumeister & Boone, 2004; Moffitt et al., 2011

[65] Duckworth & Seligman, 2005

[66] Dewall et al., 2007; Biro & Stamps, 2008; Careau et al., 2008; Biro & Stamps, 2010

[67] Watkins, 2008

[68] Freeman, 1968; Murphy et al., 1999; Cami & Farre, 2003; Tedeschi & Calhoun, 2004; Watkins & Mounds, 2004; Hughes, Alloy & Cogswell, 2008

[69] Boyer & Lienard, 2006; Baumeister et al., 2007a; Gailliot & Baumeister, 2007; Taylor & Stanton, 2007

[70] Baxter, 1989; Cohen et al., 1989; Kennedy et al., 2001; Convit, 2005; Pitel et al., 2009; Kuperman et al., 2010; Dong et al., 2012

[71] Hameroff, 2010; Schulman et al., 2009

[72] Gonzales et al., 2010

[73] Craft, Cholerton & Baker, 2013

[74] Wilkinson, 2010

[75] Miklowitz & Johnson, 2006; Buckholtz et al., 2010; Hare et al., 2010; Nestor et al., 2011

[76] Erlanson-Albertsson, 2005; Nestler, 2005; Nestor et al., 2011

[77] Ong & Muhlhausler, 2011

[78] Cawley & Meyerhoefer, 2012; Cawley & Ruhm, 2012

[79] Rosenthal, 2006

[80] Gal & Liu, 2011

[81] Hartman & Matsuno, 1992; Sharma & Annila, 2007

[82] Sun deities with their solar boats and sun chariots can be found in every civilization back to Neolithic times (Giese, 1976). And all of the modern major religions (Buddhism, Christianity, Hinduism, Judaism, Islam and Sikhism) state that light (energy) is synonymous with God (Das, 2010). In both the Old and New Testaments, light was primal, as in the creation "And God said let there be light, and there was light" (Genesis 1:3). Light was the essence of God, as in "God is light and in him is no darkness at all" (First Epistle General of John 1:15); "The Lord is my light and my salvation." (Psalms 27:1); "I am the light of the world: he that followeth me shall not walk in darkness, but shall have the light of life" (John 8, 12). God is also depicted as providing energy as in nourishment, "I am the bread of life: he that cometh to me shall never hunger" (John 6, 35).

[83] Yonge, 1854

[84] Stanley, 2009

[85] Pope Francis, 2015

[86] Posnas, 2007; Wisner, 2010

[87] Mark, 2011

[88] Mark, 2011

[89] Wolszczan, 1994

[90] Mendez, 2021; https://exoplanets.nasa.gov/exoplanet-catalog/

[91] Irwin et al., 2014

[92] Bounama, Von Bloh & Franck, 2007

[93] Estimates of the total number of stars in the Milky Way Galaxy vary widely, but 100 to 400 billion (1 to 4 x10^{11}) falls within a consensus range. (10^{11})(1 %) = 1 billion; (10^{11})(0.002%) = 2 million

[94] Irwin & Schulze-Makuch, 2011; Schulze-Makuch & Irwin, 2019

[95] Darwin, 1859

9

Strengths, Limitations, and Challenges

The energy dissipation model of the evolutionary imperative is in our view a legitimate scientific theory with strong explanatory power. Like any theory, however, it needs to be tested by critical analysis against existing empirical data, and where possible, by experimentation to fully define its reach and limitations. To the extent that it has predictive value, the challenges for our biosphere, and certainly for the human species that it predicts are grave. Our final chore will be to suggest an alternative to the fate to which the model would appear to consign us.

9.1 Strengths of the Model

A scientific model becomes a theory if it satisfies three criteria: falsifiability, comprehensiveness, and parsimony. To be falsifiable, the theory must yield testable predictions. This also guarantees that the theory will eventually be required to change—by modifying it to accommodate new facts, by extending it in new directions, by adding limitations to its application, or else by rejecting it altogether.

In this sense no theory can ever be finally proven because new facts may always come to light.

The present model is falsifiable in the sense that the direction of predicted effects are known—energy will be dissipated as it is consumed—and because specific measures are available to rigorously measure the effects (energy rate density)[1] and the efficiency or effectiveness of those effects (metabolized energy dissipated per unit of work done/per expected outcome completed).[2] We have presented many examples of our model's application to observed effects. We have argued that our model fits existing data and resolves pre-existing dilemmas. However, at this point what is missing is the application of these measures to the myriad of other possible examples of energy dissipation associated with evolutionary changes. Any evolutionary process shown to occur over the long run contrary to the SLT or the PLA will bring the validity of the model into question.

Elements of the current model, especially with regard to the role of thermodynamics, can be recognized in the work of many others.[3] Our inspiration for the energy dissipation model of the evolutionary imperative derives especially from the ideas of Eric Chaisson[4] and from the mathematical analyses of Arto Annila and his colleagues.[5] The observations of Neubauer[6] and Hoelzer[7] have been particularly helpful in our attempt to formulate a more accessible version of the current model. Their work, however, was done without citing the work of Annila or reference to the PLA. Consequently, the work of others cited in support of the model has been tangential and has not involved direct tests of the model. However, in many cases others have reported findings congruent with the conclusions of Annila's group. Examples include the findings of Karl Friston[8] at University College London, James Brown[9] at the University of New Mexico, Aljoscha Neubauer[10] at the University of Graz, and Richard Wrangham[11] at Harvard University.

A theory should also be comprehensive. While a good model may accurately predict outcomes within a limited domain, to become accepted as a theory it must be applicable more broadly. Our model rests on universal scientific principles—the Second Law of Thermodynamics (SLT) and the Principle of Least Action (PLA)—that are congruent with the pillars of modern natural science: quantum theory, relativity, electromagnetic interactions, and evolution. It brings together predictions about changes in the physical world of astronomy, physics and chemistry, the biological world, and the socio-cultural world. By explaining the imperative for change at all levels, it surpasses the power of explanations based merely on complexity theory or hierarchy.

Finally, a theory should be parsimonious. Such theories are termed "elegant" because they require the fewest number of novel assumptions, and therefore provide the most economical explanations. The present model rests only on the assumptions of well-founded pre-existing laws (SLT and PLA). Save for the added element of inheritance necessary for the operation of natural selection in the living world, no additional assumptions are required. Indeed, the economy of the present model extends to a bridge from energy to information. Independent replication, validation and application of the theory by others constitute information dissipation. Critically, experimental verification has been reported for Landauer's Principle which links energy dissipation and information dissipation. A small amount of heat (energy) is released when a small amount of information—a "bit" of data is erased.[12] Information erasure cannot happen without an increase in energy dissipation.[13]

9.2 Limitations of the Model

While applications of the current model are readily demonstrable, parsimonious, and applicable over all known domains with no known exceptions, because the model involves a causal chain modified by recurring feedback rather than a unidirectional cause and effect process, the evolving outcome in any particular situation cannot be predicted explicitly.[14]

As with any attempt to explain a range of phenomena as vast as the "evolution of everything" on the basis of a single, overarching principle, a skeptical view that such an attempt is overambitious is understandable. Critics may fairly question whether evolution is ultimately driven *only* by the SLT and PLA. To be sure, evolution is manifest through a variety of proximate factors—like radioactive decay, continental drift, and energy gradients in the physical world; mutation and recombination, natural and behavioral selection, and cognitive biases in the living world. We believe the case can be made, however, that irrespective of short-term factors and influences which certainly do factor into the details of evolutionary trajectories, the ultimate driver of those trajectories is the pressure to dissipate energy in accordance with the SLT and PLA.

Clearly what is needed are direct tests of the model by independent groups, employing alternative hypotheses in the abiotic, biological and behavioral realms. Critically, assessment of energy transduction effectiveness is needed in ecological evaluations of the relative fitness of species. The measurement of optimization or effectiveness of energy transformation is especially difficult across different domains. In addition, there is the issue of to what extent effectiveness of energy dissipation involves the value of the cost of securing and utilizing the energy as opposed to the value of payoffs or benefits. How should costs be weighed against benefits in arriving at a net value? Then there is the fundamental question of how to compare energy dissipation in various realms. The rate density measure of energy flow of Chaisson[15] and the metabolic measures of Brown[16] are candidates. Finally, can some sort of depreciation measures be used to assess the costs of dwindling resources, species extinction, accumulating pollution, increasing income disparity and population dissatisfaction?

Recently, brain structures have been identified that assess the cost of effort against reward value. The release of dopamine by the brain signals cost-benefit analysis;[17] and brain functions have been related to the individual's analysis of optimality of planned actions.[18] Such brain functions would be expected if we have indeed evolved to optimize energy dissipation. Behavioral evidence of an individual's monitoring of task efficiency has also been observed in animals.[19] Economists today are grappling with the issue of the measurement of attendant costs when they are forced to include effects previously regarded as externalities in their cost/benefit analyses, such as optimal strategies for control of greenhouse gases.[20]

9.3 Implications for Human Survival

Implicit in the model is a grave message regarding the future threats to our ecology posed by resource depletion and environmental degradation. The energy limits on economic activity have been clearly delineated through the application in the Second Law to the growth of national economies, through data linking energy consumption and GDP growth, and through direct parallels observed between biological metabolism and socioeconomic change. GDP growth is entirely predictable from energy consumption.[21] Perhaps most compelling is the finding that the increase in energy consumption by human societies accounts for the otherwise anomalous decrease in birth rate in economically developed countries.[22] In agriculture-based societies prior to mechanization, the amount of energy invested in each offspring was lower, while the added manual labor provided by raising more children was an asset, enhancing family productivity. But in mechanized, technologically mature societies, a higher standard of living can be achieved, and more energy can be expended by a couple without

children than by one with children. To put it another way, humans opt to have fewer children if children do not increase one's wealth compared to earning a living without them. A lower birth rate that contributes to more effective energy consumption is exactly what the energy dissipation model of the evolutionary imperative would predict. This has profound theoretical implications because it means that optimal expenditure of energy rather than reproductive success is the proximate mechanism driving the evolutionary process, at least in humans.

The universal imperative to effectively transfer ever more energy is expressed through society's preoccupation with consumption, ignorance of conservation strategies, the quest for material wealth and status, and the universal political focus on economic growth by economists and more importantly by corporate and political leaders.[23]

Automation of the work place driven by technology is a mixed blessing. On the one hand automation increases the productivity of workers not only on the assembly line but also in primary industries such as mining, lumbering, and farming. However, many workers are losing their jobs to machines. In fact, professionals like lawyers, doctors, managers and teachers are now beginning to be declared redundant by the development of sophisticated communication, diagnostic, and decision-making software.[24]

Unsustainable growth has many manifestations that threaten collapse of the system, including species debt, ecological debt, natural debt, greenhouse or carbon debt, social debt, the debt of individual humans, as well as the more traditional measures of economic debt of consumers, corporations, and sovereign nations. The present model clearly identifies the seriousness of this fundamental anthropogenic threat to our survival.[25]

Given the inexorable forces driving us to consume ever more, feasible solutions are unlikely to involve adoption of "slow growth" solutions, "small is beautiful" solutions, utopian scenarios of voluntarily regulating increases in income disparity,[26] slowing of global population growth through government directives,[27] or technical innovations like fuel-efficient cars that increase efficiency by reducing fuel requirements. Increased efficiencies are generally not effective because they promote increased consumption as a result of reduced costs—the classic rebound effect first described by William Jevons for the coal industry in Great Britain in 1866.[28] Significant developments for reducing the environmental burden are more likely to involve the discovery of nonpolluting sustainable sources of energy, including wind, solar, geothermal, tidal, large scale hydro, hydrogen from artificial leaves, *in vitro* meat production, and energy crops like switch grass and algae.[29] Also needed are more efficient forms of long distance transmission of energy and energy storage. However, while such innovations have reduced environmental pollution, they have also reduced energy prices and increased consumption of energy.[30]

The chronic incurring of energy debt is not just a human problem. Its most consequential form in the animal world is the crash of populations whose demands exceed their natural resources.[31]

Assuming that the SLT and PLA are driving the proclivity to overconsumption and indebtedness, it is appropriate to ask whether there are any examples in the animal world in which energy expenditure is restrained and conservation is practiced in periods of surplus. The best examples are caching and food storage by many insects, birds, and mammals. Storage of energy by the individual as body fat has also been recognized as an evolved safeguard against privation in infant birds, hibernating mammals, and in humans.[32]

The obvious historically recurrent examples in human culture of resource husbanding is the evolutionary transition from the hunting of wild game and gathering of wild plants to the domestication of animals and the storage of harvested crops.[33] These innovations permitted significant increases in human populations and in the energy consumed per person.[34] However, while hunter-gatherer societies are typically egalitarian, societies which store food are prone to social inequality.[35] Indeed inequality has been suggested as a contributor to the collapse of organized societies.[36] This principle has been extended to modern developed societies in studies showing that the returns on invested capital wealth accumulate faster than the economic growth of the countries involved, thus exacerbating chronic inequality and reducing equality of opportunity.[37]

In conclusion, these cases of debt and deficit control have evolved in spite of the suzerainty of the SLT and PLA. It is apparent that while the consequences of the SLT are being restrained to a degree in the short term by conservation, the individual's life time consumption of energy, on average, continues to grow.

9.4 Alternative End Games

To this point, our argument would seem to be leading to the conclusion that growing resource scarcity and total degradation of our habitat—indeed, of the entire planet—is inexorable. Entropy would appear to be heading toward a maximum, as required by the SLT under the relentless pressure of the PLA, to a point where our planet no longer is livable. Whether this really is the only, ultimate end game of our world is therefore a fair question. In our opinion, it doesn't have to be.

9.4.1 Switching from material to information consumption

The uniqueness of the human species, as argued previously, clearly lies in our superior effectiveness at creating, verifying, storing, retrieving, and exchanging information. This fact could well be our salvation. The key is whether we are able to switch from the intensive consumption of material energy to the optimal manipulation of information as the endpoint of our endeavors as a species. Can we come to substitute acquisition and sharing of knowledge and acumen for competition to enrich our material wealth? We believe there is reason to be hopeful.

What precedents for developing an information-focused society are there? Historical examples include the life style of people adopting the monastic life in medieval times and of some people who currently espouse Buddhist ethics. There have also been cases of people who were forced by circumstances to live lives of relative penury. In other cases, the culture itself has promoted an alternative ethic to the accumulation of material wealth. For over a thousand years, Koreans lived an agricultural lifestyle based on Confucian precepts that emphasized collective management of crops with a strong ethic of caring, sharing and cooperation. Competitive aspirations were focused on achieving an idealized state of personal tranquility and oneness with others and the universe. Though material privation was typical of these cases, we believe that it need not be so. Channeling the competition for material wealth into a search for wisdom would certainly be congruent with our natural competitive proclivity. It will feel good or "natural" to engage in this endeavor, and will be gratifying to share with and assist others in their pursuit of knowledge.

Through our own experience as teachers in modern society, we have seen the value and felt the reward of playing a part in the transmission of knowledge. Teaching is a two-way endeavor that not only broadens the world and increases understanding for the student, but clarifies and deepens the insight of the instructor. From the poorest village in sub-Saharan Africa to the most affluent high-tech community in the First World, becoming educated is among the highest aspirations for all people. Hence the sharing of knowledge has become one of the highest values of contemporary societies, and often their most time-consuming activity.

One need only consider the reinforcing power of sharing reflected in the growth of the social net. People do not need to be paid to put many hours into sharing on social media. As the cultural redefinition of our goals and mores develops, our earlier primary preoccupation with the accumulation of material wealth will seem juvenile, unrefined, shortsighted, and extraordinarily selfish. It will seem increasingly natural to define personal and group status in terms of increased knowledge, aesthetic appreciation, insight, social involvement, and awareness, rather than through material wealth and possessions.

We will come to value association with those who are knowledgeable and who can clearly convey their wisdom and insight. But can humans adopt a cultural change to a new ethos that would herald such a profound shift in behavioral attitudes and morals? We think this is possible if the change does not violate the edicts of the SLT and the PLA.

In Chapter 7 we discussed the critical role of human cognition in enabling humans to cooperate in large groups. The adage that perception is reality suggests that beliefs can be very powerful in shaping human behavior. If so, the perception of increased personal wealth might lead a person to behave as if they were wealthier. Applied to birth rate this might lead one to predict that birth rate would drop if people perceived themselves as wealthier. Until recently the dangers of world population explosion have been emphasized. However new findings indicate that a rapid decline in fertility rate is occurring in many developing nations. Fertility rate is the number of children a woman is expected to have in her lifetime and is a more informative measure than simple birth rate. In Brazil the fertility rate decreased from 6.25 in 1960 to 1.81 in 2011.[38] Similar findings have been observed in parts of India and in other developing nations (Fig. 9.1).

What accounts for the rapidity of this profound change? The strongest correlate with this relatively rapid drop in fertility was not increases in wages or in services like electrification, but rather the arrival of cable television, and specifically soap opera programming. Women in the viewing population saw women on television enjoying life without raising a large number of children and believed they could do it too.[39] They espoused as possible a reality based on faith. Congruently, in Section 7.9 we described the critical role of large group adoption of a vision or mission based on faith as a transformative human cognitive capability. Status conveyed by the competitive accumulation of material wealth depends on having more than others, whereas status based on informational wealth does not. Informational wealth differs from material wealth in several ways. It conveys status only if it is shared with others. Information is not diminished by sharing; rather the person respected for their wisdom or insight earns status by being able to create and convey their wisdom to others.[40]

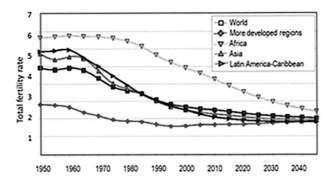

Fig. 9.1. Trends in total fertility rates from 1950 through 2050, based on data from the United Nations World Population Prospects, 2015. *Licensed under Creative Commons Attribution 3.0 Unported (https:// commons. wikimedia. org/wiki/File:Trends_in_TFR_1950-2050.png)*

The originators of information are acknowledged for being first to acquire the knowledge and for being able and willing to convey it to others. However most of us are not ideologues, and the all-wise guru is not the only model. Others recognized by our society for their knowledge include all persons engaged in providing services, those providing personal care, the healing professions, teachers in all domains, all those in the creative arts, entertainers, and professionals who guide others with their expertise. If we can adopt and apply these values successfully, those who provide us with a comfortable, predictable, sustainable, physical environment will be recognized fully as well. To illustrate this change we compare the mottos of an individual in each realm. To epitomize the life goal of the materialist we cite the motto carved into the lintel of the stone entrance of the villa of a wealthy Roman wine merchant in Pompeii (Aulus Umbricius Scaurus, 60-70 AD). The inscription was "Prodest gaudium meum" which translates as "Profit is my joy." An equivalent motto for the information age would be "Dividere intellectus gaudium meum" or "Sharing understanding is my joy."

How do we get from here to there without going through end-of-time disasters? Grant has provided a behavioral analysis of factors responsible for excessive material consumption and has described the interventions necessary to bring about societal adoption of more sustainable forms of consumption.[42] The switch from a consumer economy based on material wealth to a cultural economy based

on informational wealth will be facilitated by enactment of appropriate strategies, such as life-long educational programs, a shorter work week, and universal programs for delivering health care and retirement support. If the proportion of the working age population no longer working for a living income continues to increase—it grew from 20% to 40% in the past 40 years in the United States[41]—this will force more people into the information economy. The current growth of the low paid service economy is the beginning of this trend.[42]

Initially traditional modes of pursuing this change in values will include refocusing on local activities, like eating locally, working or volunteering in the community, taking part in neighborhood leisure and sports activities, and building local economies through strategies such as micro financing. Fortuitously, the technologies for disseminating information broadly have recently been developed that are relatively inexpensive, almost universally available and much less demanding of material energy than was previously the case, such as personalized wireless telecommunications and the internet versus airline travel, or electronic books and periodicals versus printed and shipped publications. Operation of internet cloud systems, crowdsourcing, home office businesses, and Wikinomics-type strategic endeavors are much less demanding of material resources than traditional modes of operation. Face-to-face contact is being supplanted by internet-enabled interactions through e-health, e-education, e-shopping, e-banking, e-research, e-mapping, e-tax filing, and other similar mechanisms. The socialization of the internet through cell phone applications such as texting, internet gaming sites, avatar communities, instant answer, and other internet social services such as Facebook, YouTube, Google, Twitter, Instagram, WhatsApp, WeChat, Zoom, and Skype augur for focus on life style, other than material gain. They are also a tremendous boon for isolated people: the elderly, handicapped, and remotely located. Thus, the reorientation of societal goals from material wealth to informational wealth will encourage reduced expenditure of natural wealth and will hopefully result in reduced expectancies regarding personal material requirements. The latter will likely not be a primary focus as it was in the earlier ascetic traditions of the great religions, but will happen as a natural consequence of the shift to creation and dissemination of knowledge and growing awareness and enlightenment. Parenthetically, past instances of studying the scriptures, praying, and manuscript copying can be viewed as historical exemplars of employment in a social service involving information processing with a low cost to the material world.

9.4.2 Factors influencing the shift from material to information societies

The principle threats to our success in realizing this new "informational world" will be the ongoing momentum of our material habits,[43] and the possibility that the internet's democratization will lead to inundation by a flood of information debased by its sheer volume, fabrication, frivolous content, venomous intent, and indiscriminate dissemination. The current backlash against intrusive advertising and "oversharing" on social media is a case in point.[44] Additional threats will come from an authoritarian response to perceived security threats, or the spread of proprietary corporate control in the form of restricted access and biased content. The momentum of our current material habits will ultimately lead to resource scarcity and increasing prices that will be amplified by increasing income disparity.[45]

Degradation of a democratized internet is a greater threat. Addressing it successfully will depend on several factors. One is the advantage for effective energy dissipation in cooperative versus competitive enterprise, mentioned in Section 7.9. Another factor is that unlike material wealth, the total amount of informational wealth is ever expanding and essentially unlimited.[46] The flood of disinformation and manipulation can be offset by efficient retrieval of sought-after information through high-speed search engines and reliable filters such as trusted news organizations.

A third factor alluded to in the previous section is that in contrast to material wealth, the value of a piece of information is not diminished for the sharer by sharing it with someone else.[47] Thus a shift from a material to an information economy supports a cooperative, noncompetitive bias.

This change is profound. Competition to acquire something material before the competitor acquires it reflects the exclusive nature of a lifestyle focused on accumulating material wealth. Such wealth is maintained by excluding others from possessing the goods unless they materially pay the owner. An information economy could adhere to this strategy by enforcing laws allowing the originator to exercise copyright ownership by charging others for use of the information and by not allowing them to amend it.

But achieving proprietary control of informational compared to material wealth is difficult, expensive, and counterproductive. Witness the difficulty that web-based systems like the Massive Open Online Courses are having in protecting copyrights.[48] However, as noted above in Section 9.4.1 we are already seeing examples of cooperative enterprises like Wiki, crowd sourced innovative developments and open source software like Linus and open format application development, which have the advantage of more rapid development by enlisting multiple contributors and users. The innovative advantage for producers of product modification through consumer innovation is well documented.[49] In effect we are moving toward mass-scale entrepreneurship. Because such informational innovation will be intrinsically rewarding, it will be self-sustaining. If this is coupled with a move toward a universal living wage, we will have significantly reduced the deleterious effects on resources and the environment of the natural drive to accumulate material wealth. What we are witnessing is the natural expression of switching from an energy-based to an information-based economy.

A fourth factor is that the greater degree of interconnectedness provided by advancing technology will lead to a perceived decrease in social, cultural, and racial distance from each other and to growing interdependence. Congruently, growing interconnectedness fostered by the internet is beginning to provide evidence for a switch from competitive hyper-consumption to cooperative or collaborative consumption, as in the sharing, renting, and lending of books (Bookmooch), bicycles, (Freecycle), cars (Zipcar), taxi service (Uber, Lyft), houses (Couchsurfing), money (Lendingclub), assisted living (Cohousing), and dying (Positivepalliativecare).

Evidence for movement away from the goods to the services sector can be seen in the 5-fold increase over the last half century in America's Gross Domestic Product (GDP) while the physical weight of the GDP did not change. This indicates that much of the increase was in the services sector (financial services, education, health services, and social services), as well as in electronics. Services involve innovation and exchange of information rather than of material goods. Importantly, compared to material waste, accumulation of informational waste does not threaten to pollute the environment. Taken together, these factors will reduce the likelihood that looming material shortages will result in the adoption of fear-based authoritarian policies of unrestrained competition.[50]

Unfortunately, under recessionary conditions there is a dark underside to the employment economy—also known as the gig economy, the on-demand economy, the temporary foreign worker economy, the sharing employment economy and the Uber-ized economy. Employers have shifted from full time employment with benefits (unemployment insurance, health insurance, sick pay, paid vacations, workers compensation) to short term temporary contract labor. They have done this by converting full time employment to short notice on-time employment. It is equivalent to the part time, piece work employment of the industrial era before the 1930's. So, in actuality the sharing economy offers employers lower wages and savings of up to 30%. It is rationalized by employees idealistically as offering greater choice of employment, sharing their jobs with others, personal job satisfaction, more control over what they spend their earnings on, and greater flexibility in their busy careers. The appeal is made special by the seductive technical attraction of reliance on the Internet, applications and texting on cell phones for instant messaging of the instantly employed and unemployed. The grotesque reality is that employers under cover of the lack of government regulation are able to leverage the new technology to offer jobs through anonymous interactions with online brokerages to determine who will charge least for the labor they need,

Hopefully, these debilitating arrangements can be avoided, and a viable lifestyle pursued by adopting revised ethical guidelines based on the effects of switching from a material to an informational economy. The new ethics will advocate sharing knowledge rather than amassing or consuming material goods, achieving satisfaction and recognition from contributing to the informational economy, actively creating and sharing informational innovations, growing global interdependence, and recognition of a shared responsibility for effective protection of the environment and the shepherding of resources. The urgency of the development of such a code is reinforced by the likelihood of the immanent development of forms of artificial intelligence which will need ethical codes to guide their behavior.

There is a final factor fostering a cooperative era of global integration. Historical analysis suggests that successive revisions of the major religions have embraced an ever-larger portion of humanity, eventually admitting people of all races and nations to the faith.[51] In this regard it is relevant that some branches of traditional religions have embraced aspects of the issue of future survival, such as sustainability, conservation, and pollution control, which have clear universal appeal. These organizations and their nonsectarian variants have the potential to motivate people across racial, ethnic, political, and economic boundaries to join in sustained cooperative efforts.

9.5 Future Directions

9.5.1 Research

Further research is needed to independently assess the current model's theoretical and empirical base. Formal critiques are needed of the foundational work of Annila and his colleagues.[52] Are there preferable alternative ways of interpreting their principal thesis that the SLT in concert with the PLA is able to account for all changes in energy dissipation over time, in all abiotic and biotic domains, and at all levels from the galactic to the subatomic? The independent findings of Chaisson (2001) are among the most impressive in supporting the universal expression of the SLT, even though Chaisson does not interpret them as such. Independent replication and extension of his work is needed. The same can be said of the concepts advanced by Vermeij[53] and Brown.[54]

Has the PLA legitimately been applied and interpreted by Annila's group? To our knowledge no other group has applied the PLA to biological and behavioral realms. Is this an oversight or is the PLA interpretable in another way? What exactly is the best way to define and measure least action in disparate realms?

A related issue is the search for examples of non-optimal, or inefficient manifestations of the SLT. Such cases would provide a significant check on the validity of the PLA. For example, there are many studies of tool use by animals (see Section 7.1); but to our knowledge, very few studies have formally assessed the energetic consequences of tool use by animals.[55] In the same vein, can the consequences of technological innovations which enhance efficiencies be reassessed to validate the rebound effect encapsulated in the Jevons paradox?[56]

In Section 8.4 we discussed how increased complexity, diversity, and/or hierarchy arise in the face of the SLT. While our explanation lies within the mainstream of thinking about how local decreases in entropy can occur within a larger system whose overall entropy increases, a double dissociation would be helpful, e.g. a demonstration that increased complexity and diversity do in fact *accelerate* increased entropy of an entire system. A better way to conceptualize these phenomena may be to distinguish between the proximal means and the ultimate ends of evolutionary processes. For instance, we argued in Section 9.3 that survival and reproduction involve proximal strategies for realizing the ultimate imperative of the SLT and PLA. The example of the declining birth rate of developed nations and more recently in less developed countries is a clear dissociation of the proximal productive process from the ultimate consequence of energy dissipation.

Strategically, the scientific community will be a crucial player in finding solutions to the challenge that runaway energy consumption poses to our survival. Engagement in this mission for physicists and chemists will mean recognition that the world of closed systems with its fixed parameters cannot deal with the probabilistic indeterminacy of open systems that incorporate environmental fluxes and the quasi-random variations of adapting life forms. Hopefully this will lead to the development of more robust and versatile formal models for application to the biotic world. For biologists, experimental psychologists, and economists, the ecological validity of measures and manipulations will of necessity be stretched to include environmental and long-term temporal externalities. Examples of such efforts include assessing energy consumption across levels of the ecosystem;[57] assessing human well-being in various populations;[58] and a system wide proposal for controlling nitrogen management comparable to that for controlling carbon emissions, linking the domains of food and fuel consumption.[59]

Research is needed to establish appropriate system-wide measures of energy flow. We need these measures to provide assessment of progress or lack thereof in our efforts to switch from material to information consumption. To clearly understand the magnitude of the problem and to track changes will require a theoretical framework that incorporates all the interacting factors. The field of ecological economics may provide a possible candidate for such a framework.[60] In writings published in the 1920s, Nobel Laureate Frederick Soddy conceptually replaced the then current economic model of wealth generation as a perpetual motion machine, with a thermodynamic model, which showed how to avoid the cyclic economic downturns generated by debt accumulation without regard to wealth.[61] Soddy's seminal work ushered in ecological economics. Promising elaborations of this approach include the behavioral analysis of sustainability,[62] the reinvented capitalism of Porter and Kramer,[63] the proposal for a steady-state economy by Daly,[64] and the Genuine Wealth model of Anielski.[65] The latter includes several different indicators of capital, including natural, social, and human assets, and looks beyond quarterly profits to the well-being of staff and the community, as well as future generations, and the conservation of natural resources.

9.5.2 Goals for social and political activists

In Section 9.3 we reviewed the dismal prospects for human survival if we do not recognize the threats posed by our compliance with the SLT and the PLA to create wealth and dissipate energy as rapidly as possible. We described an alternative involving a switch from a material to an information creation society. This alternative poses daunting challenges with many pitfalls, but in principle is achievable.

What lessons can be learned from the record of sustainable productivity in human history? A strong case has been made for the importance of an inclusive as opposed to an exclusive society in the prospering of a nation.[66] Historical periods in which nations have included a large portion of the population in decision making through the sharing of wealth, education, and healthcare have been characterized by greater growth and more effective adaptation to changes. During such periods in their history countries are less prone to social and economic stagnation and less prone to societal collapse when faced with challenges. In contrast, during periods in which information access is restricted, wealth is sequestered, and decision making is not shared, the quality of life declines for the majority, national productivity falls, innovation lags, and social divisions are exacerbated between the rulers and the dispossessed as well as between competing elites. Increased sharing of information and collaborative decision making not only promote increasing productivity but also the generation of optimal resolution of threats to its continuation.

Some challenges such as pollution and changing climate cross national boundaries. With these issues, the argument favoring national inclusiveness on confronting national issues, leads to the conclusion that inclusiveness should be transnational. By the same token, if the outcomes affect other species and future human generations, their interests must also be represented.

The benefits of transparency, accountability and distributed decision making, provide a strong rationale for active participation by all interests in society. In his reflections on America, De Tocqueville noted the remarkable political activism he observed in many communities.

There are many issues for political activists that are germane today, whether through self-representation or on behalf of others. A brief and partial list of suggestions includes the following:

1. Supporting legislation ensuring personal rights and freedoms
2. Supporting a fair and equitable legal system
3. Supporting measures reducing birth rate
4. Reducing material income disparity with an assured basic income, but maintaining a system that rewards the originators of informational wealth
5. Supporting universal education through the elementary and secondary levels as well as retraining programs
6. Supporting accessible health care
7. Supporting the incapacitated (handicapped, mentally ill, and elderly)
8. Securing worker and community representation on decision-making councils, whether in industry or in government
9. Growing the service economy
10. Supporting the development of standards for a steady state economy, as promulgated by the International Standards Organization
11. Supporting corporate and government transparency and accountability
12. Supporting international institutions, such as the International Court of Justice, the World Health Organization, the United Nations, and the International Monetary Fund
13. Supporting the move toward sustainable consumption of resources
14. Supporting the development and maintenance of carbon and pollution controls for water, air, and soil
15. Supporting extraterrestrial exploration.

Rather than advocating a change in values, the above programs for political action focus on directly approaching the problems of resource depletion and environmental pollution arising from global and trans-generational material growth with an inclusive social system of information sharing and distributed decision making. Once adopted, these programs could be expected to promote a change in our values—moving us toward a viable information-based economy and life style. Values are a distillation of our life styles. It would be futile to try to impose new values on an unchanged life style.

Summary of Chapter 9

The energy dissipation model of the evolutionary imperative meets the three criteria of a good theory: falsifiability, comprehensiveness, and parsimony. It is falsifiable in that the direction of predicted effects are known—energy will be dissipated optimally, and entropy will be increased—and because specific measures, such as energy rate density, are available to rigorously measure its effects. It is comprehensive in that it applies to a wide array of phenomena at every level of resolution, from the subatomic to the supragalactic. And it is parsimonious is relying on only two well-founded pre-existing theories: the SLT and PLA.

However, the model is limited by certain challenges as well. While the model is testable, in terms of currently operating processes and historical trajectories, the evolutionary outcome of any particular starting condition cannot be predicted explicitly because the model involves a causal chain modified by recurring feedback rather than a unidirectional cause and effect process. Elements of the model, especially with regard to the SLT, have been well studied by many others, but the critical role of the PLA has only been pursued in depth by Arto Annila[67] and his colleagues. While we are clearly sympathetic to the view that Annila's analysis of the seminal role played by the PLA in resolving energy gradients, and hence powering change, is correct, the extent and generalizability of those concepts need a critical assessment by others. In particular, the validity of extending the influence of the SLT and PLA to the behavioral, social, and cultural realms needs to be validated. And that in itself may call for a clearer notion of how to measure "optimality" and "least action" across many realms and multiple dimensions of space and time.

Because the SLT and PLA are fundamental properties of nature, they are not likely to be easily reversed, despite a logical recognition that the processes they are driving could eventually overwhelm the biosphere in a way that will render it unlivable. The proximal forces compelling ever more energy consumption and dispersal are powerful. While they increase the economic output of societies, they exacerbate income disparity and foster the abuses of hierarchical control. While technology enables greater energy efficiency, it invites greater energy consumption. Conservation measures in the short run can constrain the effects of the SLT and PLA, but the ultimate consequence of these dominant laws of nature appears to be an inexorable drive toward equilibrium, which equals death, to our biosphere.

Yet embedded within the SLT itself lies the potential for salvation. That potential rests on the relationship between energy and information. While energy and information can be shown to be related in a quantitative way—and thus capable of being interchanged—information is essentially infinite in scope, does not degrade the physical environment, is easy to share without loss of its effect, is difficult to keep sequestered, and has great power. Thus, if the values of human culture were to shift from a focus on the consumption of material energy to the generation and manipulation of information, the biosphere would have a greater chance of remaining stable and our longevity as a species would at least be prolonged.

Scientists and engineers will be called upon to play a critical role in restraining the destructive consequences of runaway energy consumption in the short term, and in handling the transition from an energy-consuming to an information-processing society over the long run. On everyone else falls the obligation to learn from the lessons of history, to look at the trajectory on which we are embarked, and to make adjustments now—while we can. A society that is more inclusive, transparent, collaborative, and locally dependent but internationally aware, is much more likely to restrain the evolutionary imperative, driven with increasing speed by the SLT and the PLA, toward change at a rate that the individual, society, and biosphere cannot sustain.

References and Notes

[1] Chaisson, 2009

[2] Enquist et al., 2003; Srinivasan, 2010

[3] Calvin, 1969; Wicken, 1985; Brooks & Wiley, 1988; Matsuno & Swenson, 1999; Demetrius, 2000; Morowitz, 2002; Bejan, 2010; Kleidon, 2010; Morcos et al., 2014

[4] Chaisson, 1987, 2001, 2013

[5] Annila & Annila, 2008; Jaakkola, El-Showk & Annila, 2008; Kaila & Annila, 2008; Annila, 2009; Annila & Salthe, 2010; Wurtz & Annila, 2010

[6] Neubauer, 2012

[7] Hoelzer, Smith & Pepper, 2006; Tessera & Hoelzer, 2013

[8] Friston, 2010

[9] Brown et al., 2004

[10] Neubauer & Fink, 2009

[11] Wrangham, 2013

[12] Berut et al., 2012

[13] Vaccaro & Barnett, 2011

[14] Annila, 2009; Annila & Kuismanen, 2009

[15] Chaisson, 2001

[16] Brown et al., 2004

[17] Day et al., 2010

[18] Hare et al., 2010; Rangel & Hare, 2010

[19] Biro et al., 2003; St Clair & Rutz, 2013

[20] Daly & Farley, 2004

[21] Brown et al., 2011

[22] Moses & Brown, 2003

[23] Victor, 2008

[24] Ford, 2015; Kaplan, 2015

[25] Ceballos et al, 2015; Zachos et al., 2001; Hansen et al., 2005

[26] Martin, 1999; Shankar & Shah, 2003

[27] Cohen, 1995

[28] Owen, 2010

[29] Edelman et al., 2005; Lewis & Nocera, 2006; Schmer et al., 2008; Lovelock, 2009

[30] Herring & Sorrell, 2009

[31] Stiner et al., 1999; Hallett et al., 2004

[32] Phillips & Hamer, 1999

[33] Testart, 1982; Diamond, 2002

[34] Chaisson, 2009

[35] Testart, 1982

[36] Diamond, 2005

[37] Picketty, 2014

[38] Ferrara, Chong & Duryea, 2012

[39] Jensen & Oster, 2009; Westoff & Koffman, 2011

[40] Thomas Jefferson recognized this, in a letter in 1813 to Isaac McPherson: "If nature has made any one thing less susceptible than all others of exclusive property, it is the action of the thinking power called an idea, which an individual may exclusively possess as long as he keeps it to himself; but the moment it is divulged, it forces itself into the possession of everyone, and the receiver cannot dispossess himself of it. Its peculiar character, too, is that no one possesses the less, because every other possesses the whole of it. He who receives an idea from me, receives instruction himself without lessening mine; as he who lights his taper at mine, receives light without darkening me." (https://en.wikipedia.org/wiki/Intellectual_property 2015)

[41] Picketty, 2014

[42] Autor & Dorn, 2013

[43] Grant, 2010

[44] Rose, 2011

[45] Picketty, 2014

[46] Shannon & Weaver, 1949

[47] Karnani, Pääkkönen & Annila, 2009

[48] Cheverie, 2013

[49] Tellis, Yin & Bell, 2009

[50] Janis, 1983

[51] Wright, 2000; and see Chapter 8 of this book.

[52] Sharma & Annila, 2007; Annila & Annila, 2008; Kaila & Annila, 2008; Annila & Kuismanen, 2009

[53] Vermeij, 2004

[54] Brown et al., 2004

[55] N'guessan, Ortmann & Boesch, 2009; Rutz et al., 2010

[56] Owen, 2010

[57] Enquist et al., 2003

[58] Ryan & Deci, 2001

[59] Socolow, 1999

[60] Daly & Farley, 2004

[61] Zencey, 2009

[62] Grant, 2010

[63] Porter & Kramer, 2011

[64] Daly, 2008

[65] Anielski, 2007

[66] Acemoglu & Robinson, 2012

[67] Kaila & Annila, 2008; Annila, 2017, 2021

10

Summary and Conclusions

This book describes how and why the universe has evolved, at every level of resolution. Our thesis is that the evolutionary imperative is a consequence of two fundamental laws of nature: the Second Law of Thermodynamics (SLT) and the Principle of Least Action (PLA). The resolution of energy gradients under the pressure of the PLA in the direction dictated by the SLT accounts for many effects including all of the organization of matter and energy that has ever come about; all the complexity that astronomical and geophysical forces have created in the physical world and that random variation and natural selection have induced in the living world; the diversity of everything from rocks and rivers to microbes and macro-organisms; the behavior of animals, the elements of culture, the structure of societies, the operation of economic systems, and the moral codes we live by.

Energy gradients exist because local concentrations of it (and its equivalent as mass), born of inhomogeneities in the fabric of space-time, arose in the wake of the universe as it expanded.

The fundamental forces of nature (the strong and weak forces, electromagnetism, and gravity) resolve these inhomogeneities by releasing potential energy into one of its various dynamic forms. The SLT says that this release will occur spontaneously in only one direction, namely toward one in which the system is left less organized (in a higher entropic state) and less capable of performing useful work after the transition than before. The PLA further states that the release will occur in the least time and along the straightest path possible.

At seeming odds with these physical imperatives is the observation that the universe has become more granular, complex, and diverse over evolutionary time. Stars have been born and clusters of them have formed into galaxies. Planets have condensed in orbit around the stars, and the surface of those planets have been sculpted into a great variety of worlds, depending on their starting composition, physical properties, access to mass and energy, and—since everything evolves over time—age. On some of those planets, a highly organized form of molecular matter has come alive and given rise to a mechanism for information storage that records its own history and dictates the form and function of its own self-generating perpetuation.

Thus, while the SLT would seem to predict that the world would unwind rather than complicate itself, the opposite often appears to be the case *at the local level*. Scientists have long recognized the dilemma but reconciled it with the fact that the system as a whole—whether mountain range, flowing river, living plant, or behaving animal—does over time degrade more energy than can be accounted for by the work it performs or the organization it enhances. The difference is, in fact, a net increase in entropy of the system as a whole, as required by the SLT.

The flow of energy through a system has long been known to organize that system—at least when the energy flows at a rate within optimal bounds. For it is the organization of the system that enables a more efficient consumption and degradation of energy. This is the consequence of the PLA, driving the conversion of potential energy to its dynamic form in the fastest and most complete way possible. The complex shape of a river delta is the set of channels that delivers the greatest volume of water through the allowable topography in the shortest time to the sea. The branching patterns of a tree have been guided in their growth to maximize their exposure to the sun, to convert more electromagnetic energy per unit of biomass into chemical energy through photosynthesis. The elaborate and energy-consuming mating ritual of many animals is part of the selective mechanism for ensuring reproductive success of the fittest individuals, so that more will be born, or those that are born will be more likely to survive, to consume more energy.

In recent decades, we have come to appreciate that group psychology and culture operate to promote cohesive social structures which are more effective than lone individuals or small groups at harnessing resources. Ecological theory has emphasized the operation of the interdependent web of life, extending ultimately to the level of the entire planet-as-living-organism. Economic theory has shown that economic systems follow lawful patterns, with a tendency like living organisms to consume resources for building

wealth—inevitably degrading those resources as they do so. Fundamentally, all these systems are driven by energy. This energy will ultimately be degraded into a higher state of entropy overall, even as it builds complexity and achieves great work in so doing. And because that consumption of energy for doing work and maintaining order at the local level is driven to do so in the fastest and most efficient way possible by the PLA, variations in the system, and proliferation of new systems that bring about innovative mechanisms for achieving more consumption of energy, promote a net increase in complexity and diversity over time. Thus, the tendency for the world at every level to become more complex and diverse is driven ultimately by the SLT and the PLA. The evolutionary process may on occasion transform energy by reducing rather than increasing complexity and or diversity, indicating that complexity itself is not an evolutionary drive but rather an emergent property representing the most frequently occurring optimal path for achieving energy consumption.

The purpose of this book, then, has been to examine the forces underlying changes in the natural world of physics and chemistry, and extend those forces to include the ultimate basis for evolutionary changes in biology and culture. Those forces enlist the SLT and PLA in driving every inanimate and animate system toward more optimal rates of energy transformation. This model extends to animal behavior as well as human learning and cognition. Phenomena as diverse as evolutionary fitness, animal and human conduct, genetic and behavioral variation, cognitive biases, volitional control, human culture, and the operation of governments and economies are all seen as manifestations of the imperative nature of these natural forces.

From the perspective of human livability, the accelerated pace of change and accumulation of wealth brought about by these forces has been too much of a good thing. Increased energy consumption, while helping to achieve a comfortable living for many, has begun to degrade the environment to an unsustainable degree, and has led to economic and social inequities that threaten social stability. The dominance of humans over other species, which our technology and cognitive abilities have enabled, as well as our own personal or in-group well-being relative to that of others, is due to our more effectively following the same primal imperative: energy dissipation. However, as noted in Chapter 9, this compulsive pursuit of material wealth puts us firmly on the path toward despoiling the planet and extinguishing human life in the process. Solutions that involve the slowing of growth, forsaking material gain, making do with less, redistributing income, curbing the accumulation of economic or ecological debt, reducing the birth rate, or even exerting collective will power, have few significant sustained exemplars to date. Because we are designed to pursue and enjoy the consumption of energy, those attempts which apparently have succeeded have been at such inordinate social cost, including the loss of personal freedom, that they have been terminated when the controls have been removed, as seen in the case of China's one-child policy. Thus the headlong rush to fulfill our destiny as supreme dissipaters of energy will likely result in rebalancing of accounts through traditional means—drought, famine, war, disease and population crashes—unless a more realistic and benign solution can emerge from our nature.

Human complicity with the SLT and PLA has produced a runaway catastrophe. A possible solution is for humans to switch from energy consumption to information processing as the goal for which society should strive. This would enable the human race and the world of which it is a part, to continue to function within the mandates of the Second Law (as they must) but in a more benign and beneficial manner. We could still compete as we must to live life fully, but our goal could be to accumulate a deeper understanding of ourselves and our world rather than to sequester ever more material wealth.

The foregoing account provides a troubling yet exhilarating perspective of our own history and future. It suggests that in the most fundamental way, we are part of the universe. The same laws that control the movement of the planets and stars, which fashion the physics and chemistry of watersheds and weather systems, are also prime controlling forces in the evolution of plants, animals, and ourselves. We are all principally dedicated to the dissipation of energy. The reality of this flowing tapestry of energy through the cosmos, the earth, and the life of all living things offers a transcendent view of our ultimate place in the world. Each being has its place and part in this majestic saga. Should this model of our being gain acceptance over time, this perspective of the evolutionary imperative could lead to the long-sought consilience of the sciences and other forms of knowing.[2] And hopefully, in the process, research that will increase the probability of our survival will be stimulated.

We are not alone in this cosmic tapestry; but on this planet, at least, only we have the foresight and ability to shape our ultimate contribution to the future. Our descendants' survival will depend on whether we can realize our true potential—the singular endowment of the human mind with its power of creation, verification, storage and exchange of informational energy. Ultimately, the austerity imposed by dwindling resources consumed under the edict of the Second Law will shape our destiny, one way or another.

References and Notes

[1] Greenhalgh, 2008
[2] Wilson, 1999

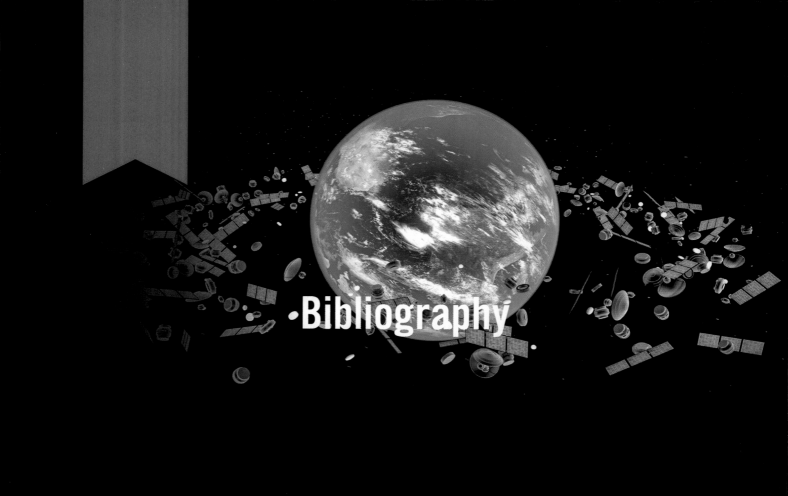

Bibliography

Acemoglu, D., and J. Robinson. 2012. *Why Nations Fail: The Origins of Power, Prosperity, and Poverty.* London: Profile Books Ltd.

Aiello, L. and P. Wheeler. 1995. The expensive tissue hypothesis: The brain and digestive system in human and primate evolution. *Curr. Anthropol.* 36:199-221.

Alberts, B., D. Bray, J. Lewis, M. Raff, K. Roberts, J.D. Watson. 1989. *Molecular Biology of the Cell.* 2nd ed. New York: Garland.

Aleshin, V.V. and N.V. Petrov. 2002. Molecular evidence of regression in evolution of metazoan. *Zh. Obshch. Biol.* 63 (3):195-208.

Anderson, P.W. 1972. More is different. *Science* 177:393-6.

Anielski, M. 2007. *The Economics of Happiness.* Gabriola Island, BC: New Society Publishers.

Annila, A. 2009. Space, time and machines. *Intl. J. Theoret. Math. Phys.* 2 (3):16-32.

Annila, A. 2017. Evolution of the universe by the principle of least action. *Physics Essays* 30:248-254.

Annila, A. 2021. Research Interests. University of Helsinki [cited 18 Mar 2021]. Available from https://www.mv.helsinki.fi/home/aannila/arto/index.html

Annila, A. and E. Annila. 2008. Why did life emerge? *Intl. J. Astrobiol.* 7:293-300.

Annila, A. and E. Kuismanen. 2009. Natural hierarchy emerges from energy dispersal. *Biosystems* 95 (3):227-33.

Annila, A. and S.N. Salthe. 2010. Physical foundations of evolutionary theory. *J. Non-Equil. Thermodyn.* 35:301-321.

Annila, A. and K. Baverstock. 2014. Genes without prominence: a reappraisal of the foundations of biology. *J. R. So.c Interface* 11 (94):20131017.

Arbib, M.A. 2008. From grasp to language: embodied concepts and the challenge of abstraction. *J. Physiol. Paris* 102:4-20.

Arbib, M.A., K. Liebal, S. Pika. 2008. Primate vocalization, gesture, and the evolution of human language. *Curr. Anthropol.* 49 (6):1053-63; discussion 1063-76.

Arnold, C. 1979. Possible evidence of domestic dog in a Paleoeskimo context. *Arctic* 32:263-5.

Arnold, L.E. 1995a. *Ablaze! The Mysterious Fires of Spontaneous Human Combustion.* New York: M. Evans and Co.

Arnold, J.E. 1995b. Transportation innovation and social complexity among maritime hunter-gatherer societies. *Amer. Anthropol.* 97:733-747.

Autor, D.H. and D. Dorn. 2013. The growth of low-skill service jobs and the polarization of the US labor market. *Am. Econ.* Rev. 103:1553-1597.

Axelrod, R. and W. D. Hamilton. 1981. The evolution of cooperation. *Science* 211:1390-6

Balu—ka, F, D.Volkmann, M. Menzel. 2005. Plant synapses: actin-based domains for cell-to-cell communication. *Trends. Plant Sci.* 10:1380-5.

Barrett, S.W. and S.F. Arno. 1982. Indian fires as an ecological influence in the Northern Rockies. *J. Forestry* 80:647-651.

Baudouin-Cornu, P. and D. Thomas. 2007. Evolutionary biology: oxygen at life's boundaries. *Nature* 445:35-6.

Baumeister, R.F. and E.J. Masicampo. 2010. Conscious thought is for facilitating social and cultural interactions: how mental simulations serve the animal-culture interface. *Psychol. Rev.* 117:945-971.

Baumeister, R.F., K.D. Vohs, C.N. DeWall, L. Zhang. 2007a. How emotion shapes behavior: feedback, anticipation, and reflection, rather than direct causation. *Pers. Soc. Psychol. Rev.* 11 (2):167-203.

Baumeister, R.F, K.D. Vohs, D.M. Tice. 2007b. The strength model of self-control. *Curr. Directions Psychol. Sci.* 16:351-5.

Baumeister, R.F., A.W. Crescioni, J.L. Alquist. 2010. Free will as advanced action control for human social life and culture. *Neuroethics* 4:1-11.

Baumeister, R.F., E.A. Sparks, T.F. Stillman, K.D. Vohs. 2008. Free will in consumer behavior: Self-control, ego depletion, and choice. *J. Consumer Psychol.* 18:4-13.

Baxter, K.L. 1989. *Energy Metabolism in Animals and Man.* Cambridge, UK: Cambridge University Press.

Beck, C.H.M., S.J. Cooper, L.J. Carter. 1994. Psychopharmacology of behavioral variability. In *Ethology and Psychopharmacology,* edited by S.J. Cooper and C.A. Hendrie. Chichester, UK: John Wiley and Sons.

Bednarik, P., K. Fehl, D. Semmann. 2014. Costs for switching partners reduce network dynamics but not cooperative behaviour. *Proc. R. Soc.* B 281:20141661.

Bednekoff, P.A. and S.L. Lima. 2002. Why are scanning patterns so variable? An overlooked question in the study of anti-predator vigilance. *J. Avian Biol.* 33:143-9.

Bejan, A. 2010. Design in nature, thermodynamics, and the constructal law: comment on "Life, hierarchy, and the thermodynamic machinery of planet Earth" by A. Kleidon. *Phys. Life Rev.* 7 (4):467-70; discussion 473-6.

Bejan, A. and J.H. Marden. 2006. Constructing animal locomotion from new thermodynamics theory. *Am. Sci.* 94:342-349.

Bejan, A. and S. Lorente. 2011. The constructal law and the evolution of design in nature. *Phys. Life Rev.* 8 (3):209-40.

Bellomo, R.V. 1994. Methods of determining early hominid behavioral activities associated with the controlled use of fire at FxJj 20 Main, Koobi Fora, Kenya. *J. Human Evol.* 27:173-195.

Berggren, J. L., and W.R. Knorr. 2014. Mathematics. *Encyclopedia Britannica Online,* http://www.britannica.com/EBchecked/topic/369194/mathematics.

Bern, Z., L.J. Dixon, D.A. Kosower. 2012. Loops, trees and the search for new physics. *Sci. Am.* 306 (5):34-41.

Berut, A., A. Arakelyan, A. Petrosyan, S. Ciliberto, R. Dillenschneider, E. Lutz. 2012. Experimental verification of Landauer's principle linking information and thermodynamics. *Nature* 483:187-9.

Bethlehem, R.A.I., S. Baron-Cohen, J. Van Honk, B. Auyeung, P.A. Bos. 2014. The oxytocin paradox. Front. *Behav. Neurosci.* 8:1-5.

Biro, D., N. Inoue-Nakamura, R. Tonooka, G. Yamakoshi, C. Sousa, T. Matsuzawa. 2003. Cultural innovation and transmission of tool use in wild chimpanzees: evidence from field experiments. *Anim. Cognit.* 6 (4):213-223.

Biro, P.A., and B. Adriaenssens. 2013. Predictability as a personality trait: consistent differences in intraindividual behavioral variation. *Am. Nat.* 182 (5):621-9.

Biro, P.A. and J.A. Stamps. 2008. Are animal personality traits linked to life-history productivity? *Trends Ecol. Evol.* 23 (7):361-8.

Biro, P.A., and J.A. Stamps. 2010. Do consistent individual differences in metabolic rate promote consistent individual differences in behavior? *Trends Ecol. Evol.* 25 (11):653-9.

Borth, D.E. 2013. Mobile telephone. *Encyclopedia Britannica Online*, http://www.britannica.com/EBchecked/topic/1482373/mobile-telephone.

Bounama, C., W. Von Bloh, S. Franck. 2007. How rare is complex life in the milky way? *Astrobiology* 7 (5):745-755.

Bowles, S. 2011. Cultivation of cereals by the first farmers was not more productive than foraging. *Proc. Natl. Acad. Sci.* USA 108:4760-5.

Boyer, P. and P. Lienard. 2006. Why ritualized behavior? Precaution systems and action parsing in developmental, pathological and cultural rituals *Behav. Brain Sci.* 26:595-613.

Breland, K. and M. Breland. 1961. The misbehavior of organisms. *Am. Psychol.* 16:681-4.

Brembs, B. 2011. Towards a scientific concept of free will as a biological trait: spontaneous actions and decision-making in invertebrates. *Proc. Roy. Soc. B Biol. Sci.* 278:930-9.

Brembs, B., F.D. Lorenzetti, F.D. Reyes, D.A. Baxter, J.H. Byrne. 2002. Operant reward learning in Aplysia: neuronal correlates and mechanisms. *Science* 296:1706-9.

Brooks, D.R., and E.O. Wiley. 1988. *Evolution as Entropy: Toward a Unified Theory of Biology.* 1st ed, *Science and Its Conceptual Foundations.* Chicago: University of Chicago Press.

Brown, J.H, J.F Gillooly, A.P Allen, V.M Savage, G.B. West. 2004. Toward a metabolic theory of ecology. *Ecology* 85 (7):1771-1789.

Brown, J.H., W.R. Burnside, A.D Davidson, J.P Delong, W.C Dunn, M.J Hamilton, J.C. Nekola, J.G Okie, N. Mercado-Silva, W.H. Woodruff, W. Zuo. 2011. Energetic limits to economic growth. *BioScience* 61:19-26.

Bshary, R. and D. Schaffer. 2002. Choosy reef fish cleaner fish that provide high-quality service. *Anim. Behav.* 63:557-564.

Bshary, R. and A.S. Grutter. 2002. Asymmetric cheating opportunities and partner control in a cleaner fish mutualism. *Anim. Behav.* 63:547-555.

Buckholtz, J.W., M.T. Treadway, R.L. Cowan, N.D. Woodward, S.D. Benning, R. Li, M.S. Ansari, R.M. Baldwin, A.N. Schwartzman, E.S. Shelby, C.E. Smith, D. Cole, R.M. Kessler, D.H. Zald. 2010. Mesolimbic dopamine reward system hypersensitivity in individuals with psychopathic traits. *Nat. Neurosci.* 13 (4):419-21.

Budd, P. and T. Taylor. 1995. The faerie smith meets the bronze industry: magic versus science in the interpretation of prehistoric metal-making. *World Archeol.* 27:133-143.

Bullock, T.H. 1977. Survey of animal groups. In *Introduction to Nervous Systems*, edited by T. Bullock. San Francisco: W.H. Freeman Co.

Burkart, J.M. and C.P. Van Schaik. 2010. Cognitive consequences of cooperative breeding in primate. *Anim. Cognit.* 13:1-19.

Burkart, J.M, O. Allon, F. Amici, C. Fichtel, C. Finkenwirth, A. Heschl, J. Huber, K. Isler, Z.K. Kosonen, E. Martins, E.J. Meulman, R. Richiger, K. Rueth, B. Spillmann, S. Wiesendanger, C.P. Van Schaik. 2014. The evolutionary origin of human hyper-cooperation. *Nat. Commun.* 5:4747

Butler, A.B. and W. Hodos. 1996. *Comparative Vertebrate Neuroanatomy: Evolution and Adaptation.* 1st ed. New York: John Wiley.

Cairns-Smith, A.G. 1982. *Genetic Takeover.* London: Cambridge University Press.

Call, J. and M. Carpenter. 2001. Do apes and children know what they have seen? *Anim. Cognit.* 4:207-220.

Calvin, M. 1969. *Chemical Evolution: Molecular Evolution Towards the Origin of Living Systems on the Earth and Elsewhere.* New York: Oxford University Press.

Cami, J., and M. Farre. 2003. Drug addiction. *New Engl. J. Med.* 349 (10):975-86.

Campbell, D. T. 1960. Blind variation and selective retention in creative thought as in other knowledge processes. *Psychol. Rev.* 67:380-400.

Campbell, N.A. 1996. *Biology.* 4th ed. Menlo Park, CA: Benjamin/Cummings.

Careau, V., D. Thomas, M.M. Humphries, D. Réale. 2008. Energy metabolism and animal personality. *Oikos* 117 (5):641-653.

Carmody, R.N., G.S. Weintraub, R.W. Wrangham. 2011. Energetic consequences of thermal and nonthermal food processing. *Proc. Natl. Acad. Sci. USA* 108:19199-19203.

Carmody, R.N. and R.W. Wrangham. 2009. The energetic significance of cooking. *J. Hum. Evol.* 57 (4):379-391.

Cawley, J. and C. Meyerhoefer. 2012. The medical care costs of obesity: an instrumental variables approach. *J. Health Econ.* 31:219-230.

Cawley, J. and C. Ruhm. 2012. The economics of risky health behaviors. In *Handbook of Health Economics,* edited by T.G. McGuire, M.V. Pauly, P.B. Barros. New York: Elsevier.

Chaisson, E. 1987. *The Life Era: Cosmic Selection and Conscious Evolution.* Authors Guild Backprint.com, Boston: Atlantic Monthly Press.

Chaisson, E. 2001. *Cosmic Evolution: The Rise of Complexity in Nature.* Authors Guild Backprint.com, Boston: Harvard University Press.

Chaisson, E. 2009. Exobiology and complexity. In *Encyclopedia of Complexity and Systems Science,* edited by R. Myers. Berlin: Springer.

Chaisson, E. 2010. Energy rate density as a complexity metric and evolutionary driver. *Complexity* 16:27-40.

Chaisson, E. 2013. Using complexity science to search for unity in the natural sciences. In *Complexity and the Arrow of Time,* edited by C.H. Lineweaver, P.C.W. Davies, M. Ruse. Cambridge: Cambridge University Press.

Chang, P.-C. 2014. The history of building construction. *Encyclopedia Britannica Online (http://www.britannica.com/EBchecked/topic/567232/Stone-Age).*

Cheverie, J. *Copyright challenges in a MOOC environment* 2013 [cited 1 Dec 2015. Available from http://www.educause.edu/ir/library/pdf/PUB9014.pdf].

Chiavassa, T., F. Borget, J.-P. Aycard, E. Dartois, L. D'Hendecourt. 2005. La chimie des glaces interstellaires: à la recherche des molecules du vivant? [Interstellar ices chemistry: the search for the molecules of life?] *L'Actualité Chimique* 283:12-18.

Clark, J.D., and J.W.K. Harris. 2005. Fire and its roles in early hominid life ways. *African Archeolog. Rev.* 3:3-27.

Cockell, C.S., and P.A. Bland. 2005. The evolutionary and ecological benefits of asteroid and comet impacts. *Trends Ecol. Evol.* 20 (4):175-179.

Cohen, J.E. 1995. Population growth and earth's human carrying capacity. *Science* 269:341-6.

Cohen, R.M., W.E. Semple, M. Gross, T.E. Nordahl, A.C. King, D. Pickar, R.M. Post. 1989. Evidence for common alterations in cerebral glucose metabolism in major affective disorders and schizophrenia. *Neuropsychopharmacology* 2 (4):241-54.

Connan, J. 1999. Use and trade of bitumen in antiquity and prehistory: molecular archaeology reveals secrets of past civilizations. *Phil. Trans. R. Soc. Lond.* B 354 (1379):33-50

Convit, A. 2005. Links between cognitive impairment in insulin resistance: an explanatory model. *Neurobiol. Aging* 26 Suppl 1:31-5.

Copley, S.D., E. Smith, and H.J. Morowitz. 2007. The origin of the RNA world: co-evolution of genes and metabolism. *Bioorg. Chem.* 35 (6):430-443.

Corballis, M.C. 2004. The origins of modernity: was autonomous speech the critical factor? *Psychol. Rev.* 111 (2):543-552.

Cowen, R. 1995. *History of Life.* 2nd ed. Boston: Blackwell.

Craft, S., B. Cholerton, L.D. Baker. 2013. Insulin and Alzheimer's disease: untangling the web. *J. Alzheimers Dis.* 33 Suppl 1:S263-275.

Crockford, C., R.M. Wittig, K. Langergraber, T.E. Ziegler, K. Zuberbuhler, and T. Deschner. 2013. Urinary oxytocin and social bonding in related and unrelated wild chimpanzees. *Proc. Roy. Soc. B Biol. Sci.* 280 (1755):20122765.

Croxson, P.L., M.E. Walton, J.X. O'Reilly, T.E. Behrens, and M.F. Rushworth. 2009. Effort-based cost-benefit valuation and the human brain. *J. Neurosci.* 29:4531-4541.

Csikszentmihalyi, M. 1991. *Flow.* New York, NY: Harper & Row.

Culver, D.C, and B. Sket. 2000. Hotspots of subterranean biodiversity in caves and wells. *J. Cave Karst Stud.* 62 (1):11-17.

Daggett, C.N. 2019. *The Birth of Energy: Fossil Fuels, Thermodynamics, and the Politics of Work.* Durham, NC: Duke University Press.

Daly, H.E. and J. Farley. 2004. *Ecological Economics: Principles and Applications.* Washington, DC: Island Press.

Daly, H.E. 2008. *A Steady-State Economy: A Failed Growth Economy and a Steady-State Economy Are Not the Same Thing; They Are the Very Different Alternatives We Face.* London: UK Sustainable Development Commission.

Damer, B. and D. Deamer. 2020. *Astrobiology* 20:429-452.

Daniels, F. and R.A. Alberty. 1961. *Physical Chemistry.* 2nd ed. New York: John Wiley & Sons.

Darwin, C. 1845. *Voyage of the Beagle.* Harvard Classics ed. New York: P. F. Collier & Son.

Darwin, C. 1859. *On the Origin of Species by Means of Natural Selection, or the Preservation of Favoured Races in the Struggle for Life.* Heritage Classics ed. New York: The Heritage Press.

Das, L.S. 2010. *The Light of God Almighty Within* 2010 [cited January 23 2016]. Available from http://www.adishakti.org/his_light_within.htm

Day, J.J., J.L. Jones, R.M. Wightman, and R.M. Carelli. 2010. Phasic nucleus accumbens dopamine release encodes effort- and delay-related costs. *Biol. Psychiatry* 68 (3):306-9.

De Moraes, C.M., W.J. Lewis, P.W. Pare, H.T. Alborn, and J.H. Tumlinson. 1998. Herbivore-infested plants selectively attract parasitoids. *Nature Gen.* 393:570-573.

De Moraes, C.M., M.C. Mescher, and J.H. Tumlinson. 2001. Caterpillar-induced nocturnal plant volatiles repel conspecific females. *Nature Gen.* 410:577-580.

deMenocal, P.B. 2014. Climate shocks. *Sci Am* 311 (3):48-53.

Demetrius, L. 2000. Thermodynamics and evolution. *J. Theor. Biol.* 206 (1):1-16.

Dennis, M.A. 2014. Internet. *Encyclopedia Britannica Online,* http://www.britannica.com/EBchecked/topic/291494/Internet.

Devenport, L.D., R.L. Hale, and J.A. Stidham. 1988. Sampling behavior in the radial maze and operant chamber: role of the hippocampus and prefrontal area. *Behav. Neurosci.* 102 (4):489-498.

Devenport, L.D. 1983. Spontaneous behavior: Inferences from neuroscience. In *Animal Cognition and Behavior,* edited by R.L. Mellgren. Amsterdam, Netherlands: North Holland.

Dewall, C.N., R.F. Baumeister, T.F. Stillman, and M.T. Gailliot. 2007. Violence restrained: Effects of self-regulation and its depletion on aggression. *J. Exp. Soc. Psychol.* 43:62-76.

Diamond, J. 2005. *Collapse: How Societies Choose to Fail or Succeed.* New York: Viking.

Diamond, J. 2002. Evolution, consequences and future of plant and animal domestication. *Nature* 418:700-7.

Dinwiddie, R. 2008. The beginning and end of the universe. In *Universe,* edited by P. Frances. New York: DK Publishing.

Dismukes, G.C., V.V. Klimov, S.V. Baranov, Y.N. Kozlov, J. DasGupta, and A. Tyryshkin. 2001. The origin of atmospheric oxygen on Earth: the innovation of oxygenic photosynthesis. *Proc. Natl. Acad. Sci. USA* 98 (5):2170-5.

Dissanayake, E. 1988. *What Is Art For?* Washington, DC: Washington University Press.

Dong, J.Y., Y.H. Zhang, J. Tong, and L.Q. Qin. 2012. Depression and risk of stroke: a meta-analysis of prospective studies. *Stroke* 43 (1):32-7.

Douglas, P.H. , and L.R. Moscovice. 2015. Pointing and pantomime in wild apes? Female bonobos use referential and iconic gestures to request genito-genital rubbing. *Sci. Rep.* DOI: 10.1038/srep13999.

Drake, G.W.F. 2012. Thermodynamics. http://www.britannica.com/EBchecked/topic/591572/thermodynamics#ref510462.

Duckworth, A.L., and M.E. Seligman. 2005. Self-discipline outdoes IQ in predicting academic performance of adolescents. *Psychol Sci* 16 (12):939-44.

Dudley, S.A., and A.L. File. 2007. Kin recognition in an annual plant. *Biol. Lett.* 3 (4):435-8.

Dunbar, R.I. 1998. *Grooming, Gossip and the Evolution of Language.* Cambridge, MA: Harvard University Press.

Dunbar, R.I.M. 1993. Coevolution of neocortical size, group size and language in humans. *Behav. Brain Sci.* 16: 681-694.

Dunbar, R.I., and S. Shultz. 2007. Evolution in the social brain. *Science* 317:1344-7.

Eastwood, M. 2003. *Principles of Human Nutrition.* Oxford, UK: Blackwell.

Eckhardt, R.B. 2006. The evolution of language: Present behavioral evidence for past genetic programming in the human lineage. *Behav. Brain Sci.* 29:284-5.

Edelman, P.D., D.C. McFarland, V.A. Mironov, and J.G. Matheny. 2005. *In vitro*-cultured meat production. *Tissue Engineering* 11:659-662.

Editors. 2012. Matter. *Encyclopedia Britannica Online,* http://www.britannica.com/EBchecked/topic/369668/matter#ref124865.

Editors. 2012. Pierre-Louis Moreau de Maupertuis. *Encyclopædia Britannica Online.,* http://www.britannica.com/EBchecked/topic/370010/Pierre-Louis-Moreau-de-Maupertuis.

Editors. 2013. Slide rule. *Encyclopedia Britannica Online,* http://www.britannica.com/EBchecked/topic/548710/slide-rule.

Editors. 2014. Factory system. *Encyclopedia Britannica Online,* http://www.britannica.com/EBchecked/topic/689674/factory-system.

Editors. 2014. Industrial revolution. *Encyclopedia Britannica Online,* http://www.britannica.com/EBchecked/topic/287086/Industrial-Revolution.

Eldredge, N. 1985. *Time Frames: The Rethinking of Darwinian Evolution and the Theory of Punctuated Equilibrium.* New York: Simon and Schuster.

Elena, S.F., and R.E. Lenski. 2003. Evolution experiments with microorganisms: The dynamics and genetic bases of adaptation. *Nat. Rev. Genet.* 4 (6):457-469.

Elkin, A. 1964. *The Australian Aborigines.* Garden City, NY: Doubleday.

Elliott, K.H., R.E. Ricklefs, A.J. Gaston, S.A. Hatch, J.R. Speakman, and G.K. Davoren. 2013. High flight costs, but low dive costs, in auks support the biomechanical hypothesis for flightlessness in penguins. *Proc. Natl. Acad. Sci. USA* 110 (23):9380-4.

Enquist, B.J., E.P. Economo, T.E. Huxman, A.P. Allen, D.D. Ignace, and J.F. Gillooly. 2003. Scaling metabolism from organisms to ecosystems. *Nature* 423:639-642.

Erlanson-Albertsson, C. 2005. How palatable food disrupts appetite regulation. *Basic Clin. Pharmacol. Toxicol.* 97 (2):61-73.

Esposito, R.U., L.J. Porrino, T.F. Seeger, A.M. Crane, H.D. Everist, A. Pert. 1984. Changes in local cerebral glucose utilization during rewarding brain stimulation. *Proc. Natl. Acad. Sci. USA* 81 (2):635-9.

Euler, L. 1744. *Methodus inveniendi lineas curvas maximi minimive proprietate gaudentes [A method for finding curved lines enjoying properties of maximum or minimum], sive solutio problematis isoperimetrici lattissimo sensu accepti [or solution of isoperimetric problems in the broadest accepted sense].* Lausanne & Geneva: Bousquet (scanned copy from Euler Archive, Dartmouth University, at http://www.math.dartmouth.edu/~euler/pages/E065.html), .

Falik, O., Y. Mordoch, D. Ben-Natan, M. Vanunu, O. Goldstein, and A. Novoplansky. 2012. Plant responsiveness to root-root communication of stress cues. *Ann. Bot.* 110 (2):271-80.

Feinberg, A.P., and R.A. Irizarry. 2010. Evolution in health and medicine Sackler colloquium: Stochastic epigenetic variation as a driving force of development, evolutionary adaptation, and disease. *Proc. Natl. Acad. Sci.* USA 107 Suppl 1:1757-1764.

Ferrara, E.L., A. Chong, S. Duryea. 2012. Soap operas and fertility: Evidence from Brazil. *Am. Econ. J. Appl. Econ.* 4:1-31.

Finlayson, C. 2009. *The Humans Who Went Extinct.* Oxford: Oxford University Press.

Fish, J.L., and C.A. Lockwood. 2003. Dietary constraints on encephalization in primates. *Am. J. Phys. Anthropol.* 120:171-181.

Fishkis, M. 2011. Emergence of self-reproduction in cooperative chemical evolution of prebiological molecules. *Orig. Life Evol. Biosph.* 41 (3):261-275.

Flannery, K.V. 1969. In *The Domestication of Plants and Animals*, edited by P.J. Ucko and G.W. Dimbleby. London: Ducksworth.

Flatz, G. and H.W. Rotthauwe. 1973. Lactose nutrition and natural selection. *Lancet* 2:76-77.

Fodor, J. 2006. The mind-body problem. In *Theories of Mind: An Introductory Reader*, edited by M. Eckert. Lanham, MD: Rowman and Littlefield.

Fordyce, R.E., and L.G. Barnes. 1994. The evolutionary history of whales and dolphins. *Ann. Rev. Earth Planet. Sci.* 22:419-455.

Ford, M. 2015. *Rise of the Robots: Technology and the Threat of a Jobless Future.* New York: Basic Books.

Fornito, A., A. Zalesky, D.S. Bassett, D. Meunier, I. Ellison-Wright, M. Yücel, S.J. Wood, K. Shaw, J. O'Connor, D. Nertney, B.J. Mowry, C. Pantelis, and E.T. Bullmore. 2011. Genetic influences on cost-efficient organization of human cortical functional networks. *J. Neurosci.* 31:3261-3270.

Forrest, B. , and P. Gross. 2004. *Creationism's Trojan Horse.* New York: Oxford University Press.

Foundas, A.L., K. Hong, C.M. Leonard, and K.M. Heilman. 1998. Hand preference and magnetic resonance imaging asymmetries of the central sulcus. *Neuropsychiatry Neuropsychol. Behav. Neurol.* 11 (2):65-71.

Fox, S.W., and K. Dose. 1977. *Molecular Evolution and the Origin of Life.* 2nd ed. New York: Marcel Dekker.

Franchi, M. and E. Gallori. 2005. A surface-mediated origin of the RNA world: biogenic activities of clay-adsorbed RNA molecules. *Gene* 346:205-214.

Freedman, D.S. 1963. The relation of economic status to fertility. *Am. Econ. Rev.* 53:414-426.

Freeman, T. 1968. On the psychopathology of repetitive phenomena. *Brit. J. Psychiatry* 114 (514):1107-1114.

Freiberger, P.A., M.R. Swaine, and W.M. Pottenger. 2014. History of computing. *Encyclopedia Britannica Online, http://www.britannica.com/EBchecked/topic/130429/computer/235927/History-of-computing.*

Friston, K. 2010. The free-energy principle: a unified brain theory? *Nat. Rev. Neurosci.* 11 (2):127-138.

Frith, C.D., K. Friston, P.F. Liddle, and R.S. Frackowiak. 1991. Willed action and the prefrontal cortex in man: a study with PET. *Proc. Biol. Sci.* 244:241-6.

Fukusako, H. 2001. Neurochemical investigation of the schizophrenic brain by in vivo phosphorus magnetic resonance spectroscopy. *World J. Biol. Psychiat.* 2:70-82.

Gagliano, M., M. Renton, M. Depczynski, and S. Mancuso. 2014. Experience teaches plants to learn faster and forget slower in environments where it matters. *Oecologia* 175 (1):63-72.

Gailliot, M.T., and R.F. Baumeister. 2007. The physiology of willpower: linking blood glucose to self-control. *Pers. Soc. Psychol. Rev.* 11 (4):303-327.

Gailliot, M.T, R.F Baumeister, C.N. Dewall, J.K. Maner, E.A. Plant, D.M. Tice, L.E. Brewer, B.J. Schmeichel. 2007. Self-control relies on glucose as a limited energy source: Will power is more than a metaphor. *J. Personal. Soc. Psychol.* 92:325-336.

Gailliot, M.T, B.M. Peruche, A. Plant, and R.F. Baumeister. 2009. Stereotypes and prejudice in the blood: Sucrose drinks reduce prejudice and stereotyping. *J. Exp. Soc. Psychol.* 45:288–290.

Gal, D., and W. Liu. 2011. "Grapes of wrath": The angry effects of exerting self-control. *J. Consumer Res.* 38 (3): 445-458.

Gallese, V. 2007. Before and below 'theory of mind': embodied simulation and the neural correlates of social cognition. *Philos. Trans. R. Soc. B: Biol. Sci.* 362: 659-669.

Galwey, N.W. 1995. Quinoa and relatives. In *Evolution of Crop Plants*, edited by J. Smart and N.W. Simmonds.

Gardner, H. 1973. *The Arts and Human Development.* New York, NY: John Wiley and Sons.

Gavrilets, S. 2014. Models of speciation: where are we now? *J. Hered.* 105 Suppl 1:743-755.

Geary, D.C. 2010. The origin of mind. In *Evolution of brain, cognition, and general intelligence. Washington,* DC: American Psychological Association.

Gell-Mann, M. 1994. *The Quark and the Jaguar: Adventures in the Simple and the Complex.* New York: W.H. Freeman & Co.

Gentilucci, M., and M.C. Corballis. 2006. From manual gesture to speech: a gradual transition. *Neurosci. Biobehav. Rev.* 30 (7):949-960.

Giese, A.C. 1976. *The Sun, Sun Myths, and Sun Worship.* New York: Springer.

Goheen, J.R., and T.M. Palmer. 2010. Defensive plant-ants stabilize megaherbivore-driven landscape change in an African savanna. *Curr. Biol.* 20 (19):1768-1772.

Gonzales, M.M., T. Tarumi, S.C. Miles, H. Tanaka, F. Shah, and A.P. Haley. 2010. Insulin sensitivity as a mediator of the relationship between BMI and working memory-related brain activation. *Obesity (Silver Spring)* 18 (11):2131-7.

Goodwin, B. 1994. *How the Leopard Changed Its Spots: The Evolution of Complexity.* New York: Charles Scribner's Sons.

Goren-Inbar, N., N. Alperson, M.E. Kislev, O. Simchoni, Y. Melamed, A. Ben-Nun, E. Werker. 2004. Evidence of hominin control of fire at Gesher Benot Ya'aqov, Israel. *Science* 304:725-7.

Gould, S.J. 1977. Eternal metaphors of paleontology. In *Patterns of Evolution as Illustrated by the Fossil Record*, edited by A. Hallam. Amsterdam: Elsevier.

Gould, S.J. 1981. G.G. Simpson, paleontology, and the modern synthesis. In *The Evolutionary Synthesis: Perspectives on the Unification of Biology*, edited by E. Mayr and W.B. Provine. Cambridge, MA: Harvard Univ. Press.

Gould, S.J. 1996. *Full House: The Spread of Excellence from Plato to Darwin.* New York: Harmony.

Gould, S.J. 2000. The evolutionary definition of selective agency, validation of the theory of hierarchical selection, and fallacy of the selfish gene. In *Thinking about Evolution*, edited by R. S. Singh. Cambridge UK: Cambridge Univ. Press.

Grant, L.K. 2010. Sustainability: From excess to aesthetics. *Behav. Social Issues* 19:7-47.

Graves, C.J., V.I. Ros, B. Stevenson, P.D. Sniegowski, D. Brisson. 2013. Natural selection promotes antigenic evolvability. *PLoS Pathogens* 9 (11):e1003766.

Greco, L. 1997. From the Neolithic revolution to gluten intolerance: benefits and problems associated with the cultivation of wheat. *J. Ped. Gastroenterol. Nutrition* 24:14-17.

Greenberg, J. M. 2000. The secrets of stardust. *Sci. Am.* 283 (6):70-75.

Greene, J., and J. Haidt. 2002. How (and where) does moral judgment work? *Trends Cogn. Sci.* 6:517-523.

Greenhalgh, S. 2008. *Just One Child: Science and Policy in Deng's China.* Berkeley, CA: University of California Press.

Groopman, E.E., R.N. Carmody, R.W. Wrangham. 2015. Cooking increases net energy gain from a lipid-rich food. *Am. J. Phys. Anthropol.* 156:11-8.

Haldane, J.B.S. 1929. The origin of life. *Rationalist Annual* 148:3-10.

Hallett, T.B., T. Coulson, J.G. Pilkington, T.H. Clutton-Brock, J.M. Pemberton, B.T. Grenfell. 2004. Why large-scale climate indices seem to predict ecological processes better than local weather. *Nature* 430:71-5.

Hameroff, S. 2010. The "conscious pilot"-dendritic synchrony moves through the brain to mediate consciousness. *J. Biol. Phys.* 36 (1):71-93.

Hansen, J., L. Nazarenko, R. Ruedy, M. Sato, J. Willis, A. Del Genio, D. Koch, A. Lacis, K. Lo, S. Menon, T. Novakov, J. Perlwitz, G. Russell, G. A. Schmidt, and N. Tausnev. 2005. Earth's energy imbalance: confirmation and implications. *Science* 308:1431-5.

Harari, Y.N. 2014. *Sapiens: A brief history of humankind.* New York: Random House.

Hare, B., V. Wobber, R. Wrangham. 2012. The self-domestication hypothesis: evolution of bonobo psychology is due to selection against aggression. *Animal Behav.* 83:573-585.

Hare, T.A., C.F. Camerer, D.T. Knoepfle, A. Rangel. 2010. Value computations in ventral medial prefrontal cortex during charitable decision making incorporate input from regions involved in social cognition. *J. Neurosci.* 30:583-90.

Harris, M. 1978. *Cannibals and Kings: The Origins of Cultures.* New York: NY: Random House.

Hartenberg, R.S. 2014. Hand tool. *Encyclopedia Britannica Online,* http://www.britannica.com/EBchecked/topic/254115/hand-tool.

Hartman, H. and K. Matsuno, eds. 1992. *The Origin and Evolution of the Cell*: Singapore: World Scientific.

Hartung, F.-M., and B. Renner. 2013. Social curiosity and gossip: related but different drives of social functioning. *PLoS ONE Biol.* 8 (7): e69996.

Hawking, S.W. 1985. Arrow of time in cosmology. *Physical Rev.* D 32:2489-2495.

Hazen, R.M. 2010. Evolution of minerals. *Sci. Am.* 302 (3):58-65.

Headey, B., R. Muffels, G.G. Wagner. 2010. Long-running German panel survey shows that personal and economic choices, not just genes, matter for happiness. *Proc. Natl. Acad. Sci. USA* 107:17922-6.

Heil, M. and R. Karban. 2010. Explaining evolution of plant communication by airborne signals. *Trends Ecol. Evol.* 25 (3):137-44.

Helmer, D, L. Gourichon, H Monchot, J Peters, . M. Sana Segui. 2005. Identifying early domestic cattle from pre-pottery Neolithic sites on the Middle Euphrates using sexual dimorphism. In *First steps of animal domestication: New archaeozoological approach,* edited by J. Vigne, D. Helmer, J. Peters. Oxford: Oxbow Books.

Henderson, J. 2007. The raw materials of early glass production. *Oxford J. Archeol.* 4:267-291.

Henderson, L.J. 1913. *The Fitness of the Environment.* New York: MacMillan.

Henry, A.G., A.S. Brooks, D.R. Piperno. 2011. Microfossils in calculus demonstrate consumption of plants and cooked foods in Neanderthal diets (Shanidar III, Iraq; Spy I and II, Belgium). *Proc. Natl. Acad. Sci. USA* 108 (2):486-91.

Herring, H., and S. Sorrell, eds. 2009. *Energy Efficiency and Sustainable Consumption: The Rebound Effect,* New York: Palgrave Macmillan.

Herrmann-Pillatha, C. and S.N Salthe. 2011. Triadic conceptual structure of the maximum entropy approach to evolution. *BioSystems* 103:315-330.

Hewstone, M. 1989. *Causal Attribution: From Cognitive Processes to Collective Beliefs.* Oxford, UK: Blackwell; pp. 30-70.

Hladik, C.M., and P. Pasquet. 2002. The human adaptations to meat eating: a reappraisal. *Hum. Evol.* 17: 199-206.

Hodge, P.W. 2021. Galaxy. *Encyclopedia Britannica,* https://www.britannica.com/science/galaxy.

Hodos, W. 1986. The evolution of the brain and the nature of animal intelligence. In *Animal Intelligence: Insights into the Animal Mind,* edited by R. Hoag and L. Goldman. Washington, D.C: Smithsonian Press.

Hoelzer, G.A., E. Smith, and J.W. Pepper. 2006. On the logical relationship between natural selection and self-organization. *J. Evol. Biol.* 19:1785-1794.

Holmes, M.M., G.J. Rosen, C.L. Jordan, G.J. de Vries, B.D. Goldman, N.G. Forger. 2007. Social control of brain morphology in a eusocial mammal. *Proc. Natl. Acad. Sci. USA* 104:10548-10552.

Hughes, M.E., L.B. Alloy, A. Cogswell. 2008. Repetitive thought in psychopathology: The relation of rumination and worry to depression and anxiety symptoms. *J. Cognit. Psychother.* 22:271-288.

Hulshof, J. and C. Ponnamperuma. 1976. Prebiotic condensation reactions in an aqueous medium: a review of condensing agents. *Orig. Life* 7 (3):197-24.

Hume, D. 1739. *A Treatise of Human Nature*. London: John Noon.

Ingber, D.E. 2000. The origin of cellular life. *Bioessays* 22 (12):1160-70.

Irwin, L.N., A. Méndez, A.G. Fairén, and D. Schulze-Makuch. 2014. Assessing the possibility of biological complexity on other worlds, with an estimate of the occurrence of complex life in the Milky Way galaxy. *Challenges* 214 (5):159-174.

Irwin, L.N. and D Schulze-Makuch. 2011. *Cosmic Biology: How Life Could Evolve on Other Worlds*. New York: Praxis.

Irwin, L.N. and D Schulze-Makuch. 2020. The astrobiology of alien worlds: known and unknown forms of life. *Universe* 6:130, doi.org/10.3390/universe6090130

Iwaniuk, A.N. and I.Q. Whishaw. 2000. On the origin of skilled forelimb movements. *Trends Neurosci.* 23 (8):372-6.

Jaakkola, S. , V. Sharma, A. Annila. 2009. Cause of chirality consensus. *Curr. Chem. Biol.* 2:53-8.

Jaakkola, S., S. El-Showk, A. Annila. 2008. The driving force behind genomic diversity. *Biophys, Chem*, 134:232-238.

Janis, I.L. 1983. Groupthink. In *Small Groups and Social Interaction*, edited by H.H. Blumberg, A.P. Hare, V. Kent, M.F. Davis. New York: Wiley.

Jensen, R. and E. Oster. 2009. The Power of TV: Cable television and women's status in India. *Quart. J. Econ.* 124:1057-1094.

Jerison, H.J. 1973. *Evolution of the Brain and Intelligence*. London: Academic Press.

Jiang, P., J. Josue, X. Li, D. Glaser, W. Li, J.G. Brand, R.F. Margolskee, D.R. Reed, and G.K. Beauchamp. 2012. Major taste loss in carnivorous mammals. *Proc. Natl. Acad. Sci. USA* 109 (13):4956-4961.

Kahneman, D. 2003. A perspective on judgment and choice: mapping bounded rationality. *Am. Psychol.* 58 (9):697-720.

Kaila, V.R.I., and A. Annila. 2008. Natural selection for least action. *Proc. Royal Soc A* 464:3055-3070.

Kaplan, H.S., M. Guerven, and J.B. Lancaster. 2007. Brain evolution and the human adaptive complex. In *The Evolution of Mind*, edited by S. Gangestad and J. Simpson. New York: Guilford Press.

Kaplan, J. 2015. *Humans Need Not Apply: A Guide to Wealth and Work in the Age of Artificial Intelligence*. New Haven, CT: Yale University Press.

Kappelman, John. 1996. The evolution of body mass and relative brain size in fossil hominids. *J. Hum. Evol.* 30 (3):243-276.

Karnani, M. and A. Annila. 2009. Gaia again. *Biosystems* 95 (1):82-7.

Karnani, M., K. Pääkkönen, and A. Annila. 2009. The physical character of information. *Proc. Royal Soc. A* 465:2155-2175.

Kaufmann, M. 2009. On the free energy that drove primordial anabolism. *Int. J. Mol. Sci.* 10:1853-1871.

Kaufmann, W.J. and N.F. Comins. 1996. *Discovering the Universe*. 4th ed. New York: W. H. Freeman & Co.

Keegan, J. 1993. *The History of Warfare*. London: Hutchinson.

Kennedy, S.H., K.R. Evans, S. Kruger, H.S. Mayberg, J.H. Meyer, S. McCann, A.I. Arifuzzman, S. Houle, and F.J. Vaccarino. 2001. Changes in regional brain glucose metabolism measured with positron emission tomography after paroxetine treatment of major depression. *Am. J. Psychiat.* 158 (6):899-905.

Keosian, J. 1968. *The Origin of Life*. Edited by P. Gray, *Selected Topics in Modern Biology*. New York: Reinhold.

Kirk, C.R, N. McMillan, W.A. Roberts. 2014. Rats respond for information: Metacognition in a rodent? *J. Exp. Psychol.: Animal Learn. Cognit.* 40:249-259

Kleidon, A. 2010. Life, hierarchy, and the thermodynamic machinery of planet Earth. *Phys. Life Rev.* 7 (4):424-460.

Knutson, B, G.W. Fong, C.M. Adams, D. Hommer. 2001. Dissociation of reward anticipation and outcome with event-related fMRI. *NeuroReport* 12:3683-7.

Koob, G.F. 1992. Dopamine, addiction and reward. *Seminars Neurosci.* 4:139-148.

Kuperman, Y., O. Issler, L. Regev, I. Musseri, I. Navon, A. Neufeld-Cohen, S. Gil, and A. Chen. 2010. Perifornical Urocortin-3 mediates the link between stress-induced anxiety and energy homeostasis. *Proc. Natl. Acad. Sci. USA* 107 (18):8393-8.

Kurniawan, I.T., B. Seymour, D. Talmi, W. Yoshida, N. Chater, and R.J. Dolan. 2010. Choosing to make an effort: the role of striatum in signaling physical effort of a chosen action. *J. Neurophysiol.* 104:313-321.

Kurzweil, R. 2005. *The Singularity is Near: When Humans Transcend Biology.* New York: Viking Penguin.

Lahav, N. 1994. Minerals and the origin of life: Hypotheses and experiments in heterogeneous chemistry. *Heterogeneous Chem. Rev.* 1:159-179.

Landis, F, C.R. Russell, R.L Seale, R. Wailes, and E.B. Woodruff. 2012. Energy conversion. *Encyclopedia Britannica Online*, http://www.britannica.com/EBchecked/topic/187279/energy-conversion.

Laski, M. 1961. *Ecstasy: A Study of Some Secular and Religious Experiences.* London, UK: Cresset.

Lathe, R. 2004. Fast tidal cycling and the origin of life. *Icarus* 168:18-22.

Lehrer, M. 1993. Why do bees turn back and look? *J. Comp. Physiol.* A 172:549-563.

Lewis, N.S. and D.G. Nocera. 2006. Powering the planet: Chemical challenges in solar energy utilization. *Proc. Natl. Acad. Sci. USA* 103 (43):15729-15735.

Lotka, A.J. 1922. Contribution to the Energetics of Evolution. *Proc. Natl. Acad. Sci. USA* 8 (6):147-151.

Lovelock, J.E. and L. Margulis. 1997. The Gaia hypothesis: The Earth as a living organism. In *Redeeming the Time: A Political Theology of the Environment*, edited by S. B. Scharper. New York: The Continuum International Publishing Group.

Lovelock, J. 2009. *The vanishing face of Gaia: A final warning.* New York: Basic Books.

MacLean, R.C., A. Fuentes-Hernandez, D. Grieig, L.D. Hurstr, and I. Gudelj. 2010. A mixture of "cheats" and "co-operators" can enable maximal group benefit. *PLoS Biol* 8:e1000486.

Madden, J.R., and T.H. Clutton-Brock. 2011. Experimental peripheral administration of oxytocin elevates a suite of cooperative behaviours in a wild social mammal. *Proc. Biol. Sci.* 278:1189-94.

Map of Life. 2012. Tool use in birds. http://www.mapoflife.org/topics/topic_193_Tool-use-in-birds/.

Margulis, L. 2008. *Symbiotic Planet: A New Look at Evolution*: New York: Basic Books.

Margulis, L., and J.E. Lovelock. 1974. Biological modulation of the Earth's atmosphere. *Icarus* 21 (4):471-489.

Margulis, L. and D. Sagan. 1995. *What Is Life?* New York: Simon & Schuster.

Marean, C.W. 2015. The most invasive species of all. *Sci. Am.* 313 (2):32-39.

Mark, J.J. 2011. Democritus. *Ancient History Encyclopedia*, http://www.ancient.eu.com/Democritus/.

Martin, P. 1999. Public policies, regional inequalities and growth. *J. Publ. Econ.* 73:85-101.

Martin, R.D., D.J. Chivers, A.M. Maclarnon, and C.M. Hladik. 1985. Gastrointestinal allometry in primates and other mammals. In *Size and Scaling in Primate Biology*, edited by W.L. Jungers. New York: Plenum.

Masicampo, E.J. and R.F. Baumeister. 2008. Toward a physiology of dual-process reasoning and judgment: Lemonade, willpower, and expensive rule-based analysis. *Psychol. Sci.* 19:255-260.

Matsuno, K. and R. Swenson. 1999. Thermodynamics in the present progressive mode and its role in the context of the origin of life. *Biosystems* 51 (1):53-61.

Maye, A., C.H. Hsieh, G. Sugihara, B. Brembs. 2007. Order in spontaneous behavior. *PLoS One* 2 (5):e443.

McBrearty, S., and A.S. Brooks. 2000. The revolution that wasn't: a new interpretation of the origin of modern human behavior. *J. Hum. Evol.* 39 (5):453-563.

McNally, L., S.P. Brown, and A.L. Jackson. 2012. Cooperation and the evolution of intelligence. *Proc. Biol. Sci.* 279:3027-3034.

McPherron, S.P., Z. Alemseged, C.W. Marean, J.G. Wynn, D. Reed, D. Geraads, R. Bobe, and H.A. Béarat. 2010. Evidence for stone-tool-assisted consumption of animal tissues before 3.39 million years ago at Dikika, Ethiopia. *Nature* 466:857-860.

McShea, D.W. 1991. Complexity and evolution: what everybody knows. *Biol. Philos.* 6 (3):303-324.

McShea, D.W. and R.N. Brandon. 2010. *Biology's First Law: The Tendency for Diversity and Complexity to Increase in Evolutionary Systems.* Chicago: Univ. Chicago Press.

Meguerditchian, A., M.J. Gardner, S.J. Schapiro, and W.D. Hopkins. 2012. The sound of one-hand clapping: handedness and perisylvian neural correlates of a communicative gesture in chimpanzees. *Proc. Biol. Sci.* 279:1959-1966.

Mendez, A. 2021. Exoplanets Catalogue - Planetary Habitability Laboratory: University of Puerto Rico at Arecibo. http://phl.upr.edu/

Metcalfe, J. 2008. Evolution of metacognition. In *Handbook of Metamemory and Memory*, edited by J. Dunlosky and R. Bjork. New York: Psychology Press.

Metcalfe, J. and J. Jacobs. 2009. People's study time allocation and its relation to animal foraging. *Behav. Proc.* 83:213-221.

Miklowitz, D.J. and S.L. Johnson. 2006. The psychopathology and treatment of bipolar disorder. *Ann. Rev Clin. Psychol.* 2:199-235.

Miller, S.L. 1953. A production of amino acids under possible primitive earth conditions. *Science* 117:528-9.

Mischel, W., Y. Shoda, and P.K. Peake. 1988. The nature of adolescent competencies predicted by preschool delay of gratification. *J. Personal. Soc. Psychol.* 54:687-696.

Mitchell, M. 2009. *Complexity: A Guided Tour.* Oxford, UK: Oxford University Press.

Miyakawa, S., P.C. Joshi, M.J. Gaffey, E. Gonzalez-Toril, C. Hyland, T. Ross, K. Rybij, and J.P. Ferris. 2006. Studies in the mineral and salt-catalyzed formation of RNA oligomers. *Orig. Life Evol. Biosph.* 36 (4):343-61.

Mlodinow, L., and T.A. Brun. 2013. On the relation between the psychological and thermodynamic arrows of time. *arXiv*:1310.1095v1 [cond-mat. stat-mech]

Moffitt, T.E., L. Arseneault, D. Belsky, N. Dickson, R.J. Hancox, H. Harrington, R. Houts, R. Poulton, B. W. Roberts, S. Ross, M.R. Sears, W.M. Thomson, A. Caspi. 2011. A gradient of childhood self-control predicts health, wealth, and public safety. *Proc. Natl. Acad. Sci. USA* 108 (7):2693-8.

Moll, H., and M. Tomasello. 2007. Cooperation and human cognition: the Vygotskian intelligence hypothesis. *Philos. Trans. Roy. Soc. Lond. B Biol. Sci.* 362:639-648.

Monod, J. 1971. *Chance and Necessity.* Translated by A. Wainhouse, 1st ed. New York: Alfred A. Knopf.

Morcos, F., N.P. Schafer, R.R. Cheng, J.N. Onuchic, and P.G. Wolynes. 2014. Coevolutionary information, protein folding landscapes, and the thermodynamics of natural selection. *Proc. Natl. Acad. Sci. USA* 111 (34):12408-12413.

Morowitz, H.J. 1968. *Energy Flow in Biology.* 1st ed. New York: Academic Press.

Morowitz, H.J. 2002. *The Emergence of Everything: How the World Became Complex.* 1st ed. New York: Oxford University Press.

Moses, M.E. and J.H. Brown. 2003. Allometry of human fertility and energy use. *Ecol. Lett.* 6:295-300.

Movius, Jr., H.L., R.J. Braidwood, K. Kuiper. 2014. Stone Age. *Encyclopedia Britannica Online (http://www. britannica.com/EBchecked/topic/567232/Stone-Age).*

Mowrer, O.H. and H.M. Jones. 1943. Extinction and behavioral variability as functions of effortfulness of task. *J. Exp. Psychol.* 33:369-386.

Muller, A.W.J. 1996. Hypothesis: the thermosynthesis model for the origin of life and the emergence of regulation by Ca^2+. *Essays Biochem.* 31:103-119.

Murdock, G.P. 1945. The common denominator of cultures. In *The Science of Man in the World Crisis*, edited by R. Linton. New York: Columbia University Pres.

Murphy, F.C., B.J. Sahakian, J.S. Rubinsztein, A. Michael, R.D. Rogers, T.W. Robbins, and E.S. Paykel. 1999. Emotional bias and inhibitory control processes in mania and depression. *Psychol. Med.* 29 (6):1307-21.

N'guessan, A.K, S. Ortmann, and C. Boesch. 2009. Daily energy balance and protein gain among *Pan troglodytes verus* in the Taï National Park, Côte d'Ivoire. *Intl. J. Primatol.* 30 (3):481-496.

Naveh, Z. 2006. The evolutionary significance of fire in the Mediterranean region. *Plant Ecol.* 29:199-208.

Nestler, E.J. 2005. Is there a common molecular pathway for addiction? *Nat. Neurosci.* 8 (11):1445-9.

Nestor, L., E. McCabe, J. Jones, L. Clancy, and H. Garavan. 2011. Differences in "bottom-up" and "top-down" neural activity in current and former cigarette smokers: Evidence for neural substrates which may promote nicotine abstinence through increased cognitive control. *Neuroimage* 56 (4):2258-2275.

Neubauer, A.C., and A. Fink. 2009. Intelligence and neural efficiency. *Neurosci. Biobehav. Rev.* 33 (7):1004-23.

Neubauer, R.L. 2012. *Evolution and the Emergent Self: The Rise of Complexity and Behavioral Versatility in Nature.* New York: Columbia University Press.

Neyskens, P., S. Van Eck, A. Jorissen, S. Goriely, L. Siess, and B. Plez. 2015. The temperature and chronology of heavy-element synthesis in low-mass stars. *Nature* 517:174-6.

Norris, V., C. Loutelier-Bourhis, and A. Thierry. 2012. How did metabolism and genetic replication get married? *Orig. Life Evol. Biosph.* 42 (5):487-495.

Northcutt, R.G. 1985. Brain organization in the cartilaginous fishes. In *Sensory Biology of Sharks, Skates, and Rays*, edited by E. S. Hodgson and R. F. Mathewson. Arlington, VA: Office of Naval Research.

Northcutt, R. G. 2002. Understanding vertebrate brain evolution. *Integrat. Comp. Biol.* 42 (4):743-756.

Olson, D.R. 2014. Writing. *Encyclopedia Britannica Online (http://www.britannica.com/EBchecked/topic/649670/writing).*

Ong, Z.Y., and B.S. Muhlhausler. 2011. Maternal "junk-food" feeding of rat dams alters food choices and development of the mesolimbic reward pathway in the offspring. *FASEB J.* 25 (7):2167-2179.

Oparin, A.I. 1938. *Origin of Life.* Translated by S. Margulis, 2nd ed. New York: Dover, reprinted 1953.

Orgel, L.E. 2000. Self-organizing biochemical cycles. *Proc. Natl. Acad. Sci. USA* 97:12503-12507.

Orgel, L. E. 2003. Some consequences of the RNA world hypothesis. *Orig. Life Evol. Biosph.* 33 (2):211-218.

Orgel, L.E., and R Lohrmann. 1974. Prebiotic chemistry and nucleic acid replication. *Acc. Chem. Res.* 7:368-377.

Orton, C, P. Tyers, and A. Vince. 1993. *Pottery in Archeology.* Cambridge, UK: Cambridge University Press.

Owen, D. 2010. The efficiency dilemma. *The New Yorker*, December 20 & 27, 78-85.

Page, S. and A. Neuringer. 1985. Variability as an operant. *J. Exp. Psychol.* 11:429-452.

Partridge, B.L., J. Johansson, and J. Kalish. 1983. The structure of schools of giant Bluefin tuna in Cape Cod Bay. *Environ. Biol. Fishes* 9:253-262.

Pernu, T.K. and A. Annila. 2012. Natural emergence. *Complexity* 17:44-47.

Perry, G.H., N.J. Dominy, K.G. Claw, A.S. Lee, H. Fiegler, R. Redon, J. Werner, F.A. Villanea, J.L. Mountain, R. Misra, N.P. Carter, C. Lee, and A.C. Stone. 2007. Diet and the evolution of human amylase gene copy number variation. *Nat. Genet.* 39 (10):1256-60.

Pfeiffer, J.E. 1985. *The Creative Explosion: An inquiry into the origins of art and religion.* Ithaca, NY: Cornell University Press.

Phillips, R.A. and K.C. Hamer. 1999. Lipid reserves, fasting capability and the evolution of nestling obesity in procellariiform seabirds. *Proc. Roy. Soc. Lond. B: Biol. Sci.* 266:1329-1334.

Piaget, J. 1932. *The Moral Judgment of the Child.* New York: Free Press.

Pickersgill, B. and C.B. Heiser. 1978. Origins and distribution of plants domesticated in the New World tropics. In *Origins of Agriculture*, edited by C.A. Reed. The Hague: Mouton.

Picketty, T. 2014. *Capital in the Twenty-First Century.* Cambridge, MA: Belknap Press.

Pinker, S. 2010. The cognitive niche: Coevolution of intelligence, sociality, and language. *Proc. Natl. Acad. Sci. USA* 107 (Suppl. 2):8993-8997.

Pitcher, T., A. Magurran, and I. L. Winfield. 1982. Fish in larger shoals find food faster. *Behav. Ecol. Sociobiol.* 10:149-151.

Pitel, A.L., A.M. Aupee, G. Chetelat, F. Mezenge, H. Beaunieux, V. de la Sayette, F. Viader, J.C. Baron, F. Eustache, B. Desgranges. 2009. Morphological and glucose metabolism abnormalities in alcoholic Korsakoff's syndrome: group comparisons and individual analyses. *PLoS One* 4 (11):e7748.

Pizzarello, S. 2004. Chemical evolution and meteorites: An update. *Orig. Life Evol. Biosph.* 34:25-34.

Pizzarello, S., and Y S. Huang. 2005. The deuterium enrichment of individual amino acids in carbonaceous meteorites: A case for the presolar distribution of biomolecule precursors. *Geochim. Cosmochim. Acta* 69 (3):599-605.

Pocheptsova, A., O. Amir, R. Dhar, and R.F. Baumeister. 2009. Deciding without resources: Resource depletion and choice in context. *J. Market. Res.* 46: 344-355.

Pollick, A.S., and F.B. de Waal. 2007. Ape gestures and language evolution. *Proc. Natl. Acad. Sci. USA* 104 (19):8184-9.

Pope Francis. 2015. Laudato Si, Encyclical Letter, May 24 2015. *The Holy See, papa-francesco_20150524_enciclica-laudato-si.*

Porter, M.E. and M.R. Kramer. 2011. Creating shared value: How to reinvent capitalism. *Harvard Bus. Rev.* 89: 62-77.

Posnas, P.J. 2007. Roles of religion and ethics in addressing climate change. *Ethics Sci. Environ. Politics* 200:31-49.

Poulsen, M, and J.J. Boomsma. 2005. Mutualistic fungi control crop diversity in fungus-growing ants. *Science* 307:741-744.

Powers, S.T., and Lehmann L. 2014. An evolutionary model explaining the Neolithic transition from egalitarianism to leadership and despotism. *Proc. Roy. Soc.* B 281:20141349.

Protas, M., M. Conrad, J.B. Gross, C. Tabin, and R. Borowsky. 2007. Regressive evolution in the Mexican cave tetra, *Astyanax mexicanus. Curr. Biol.* 17 (5):452-4.

Purves, W.K., G.H. Orians, H.C. Heller, and D. Sadava. 1998. *Life: The Science of Biology.* 5th ed. Sunderland, MA: Sinauer Associates.

Rangel, A., and T. Hare. 2010. Neural computations associated with goal-directed choice. *Curr. Opin. Neurobio.l* 20 (2):262-270.

Rankin, C.H, T. Abrams, R.J. Barry, S. Bhatnagar, D.F. Clayton, J. Colombo, G. Coppola, M.A. Geyer, D.L Glanzman, S. Marsland, F.K Mcsweeney, D.A Wilson, C.-F. Wum, and R.F. Thompson. 2009. Habituation revisited: An updated and revised description of the behavioral characteristics of habituation. *Neurobiol. Learn. Memory* 92:135-138.

Raup, D.M., and J.J. Sepkoski, Jr. 1982. Mass extinctions in the marine fossil record. *Science* 215:1501-3.

Reiches, M.W., P.T. Ellison, S.F. Lipson, K.C. Sharrock, E. Gardiner, L.G. Duncan. 2009. Pooled energy budget and human life history. *Am. J. Hum. Biol.* 21:421-9.

Revedin, A., B. Aranguren, R. Becattini, L. Longo, E. Marconi, M.M. Lippi, N. Skakun, A. Sinitsyn, E. Spiridonova, J. Svoboda. 2010. Thirty thousand-year-old evidence of plant food processing. *Proc. Natl. Acad. Sci. USA* 107 (44):18815-9.

Ridley, M. 2010. *The Rational Optimist: How Prosperity Evolves.* New York: Harper.

Rizzolatti, G. and M.A. Arbib. 1998. Language within our grasp. *Trends Neurosci.* 21 (5):188-94.

Roebroeks, W. and P. Villa. 2011. On the earliest evidence for habitual use of fire in Europe. *Proc. Natl. Acad. Sci. USA* 108 (13):5209-5214.

Rose, C. 2011. The security implications of ubiquitous social media. *Intl. J. Mangmt. Info. Syst.* 15:35-40.

Rosenthal, E.C. 2006. *The Era of Choice: The Ability to Choose and Its Transformation of Contemporary Life.* Cambridge, MA: MIT Press.

Roth, G., K.C. Nishikawa, D.B. Wake. 1997. Genome size, secondary simplification, and the evolution of the brain in salamanders. *Brain Behav. Evol.* 50:50-59.

Russell, M.J., and I. Kanik. 2010. Why does life start, what does it do, where will it be, and how might we find it? *J. Cosmol.* 5:1008-1039.

Rutz, C., L.A. Bluff, N. Reed, J. Troscianko, J. Newton, R. Inger, A. Kacelnik, and S. Bearhop. 2010. The ecological significance of tool use in New Caledonian crows. *Science* 329:1523-6.

Ruxton, G. D., and M. Stevens. 2015. The evolutionary ecology of decorating behaviour. *Biol. Lett.* 11 (6):20150325.

Ryan, R.M. and E.L. Deci. 2001. On happiness and human potentials: A review of research on hedonic and eudaimonic well-being. *Ann. Rev. Psychol.* 52 (1):141-166.

Salthe, S.N. 2004. The spontaneous origin of new levels in a scalar hierarchy. *Entropy* 6 (3):327-343.

Salthe, S.N. 2010. Maximum power and maximum entropy production: finalities in nature. *J. Nat. Social Philos.* 6 (Cosmos and History):114-121.

Salthe, S.N. 1985. *Evolving Hierarchical Systems: Their Structure and Representation.* New York: Columbia University Press.

Sandford, S.A. 2008. Terrestrial analysis of the organic component of comet dust. *Ann. Rev. Anal. Chem.* 1:549-78.

Santos Granero, F. 1991. *The power of love: The moral use of knowledge amongst the Ameusha of central Peru.* London, UK: Athlone Press.

Saunders, H.L. 1979. Evolutionary ecology and life-history in the deep sea. *Sarsia* 64:1-9.

Schipul, S.E., T.A. Keller, and M.A. Just. 2011. Inter-regional brain communication and its disturbance in autism. *Front. Neurosci.* 5:1-11.

Schmer, M.R., K.P. Vogel, R.B. Mitchell, and R.K Perrin. 2008. Net energy of cellulosic ethanol from switchgrass. *Proc. Natl. Acad. Sci. USA* 105 (2):464-469.

Schmidt-Nielsen, K. 1979. *Animal Physiology: Adaptation and Environment.* 2nd ed. New York: Cambridge Univ. Press.

Schott, B.H., L. Minuzzi, R.M. Krebs, D. Elmenhorst, M. Lang, O.H. Winz, C.I. Seidenbecher, H.H. Coenen, H.J. Heinze, K. Zilles, E. Duzel, A. Bauer. 2008. Mesolimbic functional magnetic resonance imaging activations during reward anticipation correlate with reward-related ventral striatal dopamine release. *J. Neurosci.* 28 (52):14311-9.

Schroedinger, E. 1944. *What is Life? The Physical Aspect of the Living Cell.* 1st ed. Cambridge, UK: Cambridge University Press

Schultz, W. 2009. Getting formal with dopamine and reward. *Neuron* 36:241-263.

Schulze-Makuch, D. 2014. 100 million planets in our galaxy may harbor complex life. *Air & Space Smithsonian,* http://www.airspacemag.com/daily-planet/100-million-planets-our-galaxy-may-harbor-complex-life-180951598/#ixzz3904nYVjj.

Schulze-Makuch, D. and D.H. Grinspoon. 2005. Biologically enhanced energy and carbon cycling on Titan? *Astrobiology* 5(4):560-567.

Schulze-Makuch, D., D.H. Grinspoon, O. Abbas, L.N. Irwin, and M.A. Bullock. 2004. A sulfur-based survival strategy for putative phototrophic life in the Venusian atmosphere. *Astrobiology* 4 (1):11-18.

Schulze-Makuch, D. and L.N. Irwin. 2008. *Life in the Universe: Expectations and Constraints (2nd ed.)* Berlin: Springer-Verlag.

Schulze-Makuch, D., & Irwin, L.N. 2018. *Life in the Universe: Expectations and Constraints (3rd ed.).* Switzerland: Springer Praxis.

Scott-Phillips, T.C. 2007. The social evolution of language and the language of social evolution. *Evol. Psychol.* 5:750-753.

Seligman, M. 2002. *Authentic Happiness.* New York: Free Press.

Selinger, J.C, S.M. O'Connor, J.D. Wong, and J.M. Donelan. 2015. Humans can continuously optimize energetic cost during walking. *Curr. Biol.* 25:1-5.

Semaw, S., P. Renne, J.W.K. Harris, C.S. Feibel, R.L. Bernor, N. Fesseha, K. Mowbray. 1997. 2.5-million-year-old stone tools from Gona, Ethiopia. *Nature* 385:333-6.

Sen, A. 2009. *The Idea of Justice*. Cambridge, MA: Belknap/Harvard University Press.

Seth, A.K., B.J. Baars, and D.B. Edelman. 2005. Criteria for consciousness in humans and other mammals. *Conscious. Cognit.* 14:119-39.

Shahan, T. A., and P. N. Chase. 2002. Novelty, stimulus control, and operant variability. *Behav. Anal.* 25:175-190.

Shankar, R., and A. Shah. 2003. Bridging the economic divide within countries: A scorecard on the performance of regional policies in reducing regional income disparities. *World Development* 31 (8):1421-1441.

Shannon, C.E. and W. Weaver. 1949. *The Mathematical Theory of Communication*. Urbana, IL: University of Illinois Press.

Sharma, V. and A. Annila. 2007. Natural process — natural selection. *Biophys. Chem.* 127:123-128.

Shipman, P. 2014. How do you kill 86 mammoths? Taphonomic investigations of mammoth megasites. *Quaternary Intl.*, dx.doi.org/10.1016/j.quaint.2014.04.048.

Sidor, C.A. 2001. Simplification as a trend in synapsid cranial evolution. *Evolution 55* (7):1419-1442.

Signor, P.W. 1994. Biodiversity through geological time. *Am. Zool.* 34:23-32

Silk, J.B, S.C. Alberts, J. Altmann. 2003. Social bonds of female baboons enhance infant survival. *Science* 302:1231-4.

Silva, F.J., A. Latorre, and A. Moya. 2001. Genome size reduction through multiple events of gene disintegration in Buchnera APS. *Trends Genet.* 17:615-8.

Simonson, I. 2005. In defense of consciousness: The role of conscious and unconscious inputs in consumer choice. *J. Consumer Psychol.* 5:211-218.

Smaers, J.B., J. Steele, C.R. Case, and K. Amunts. 2013. Laterality and the evolution of the prefronto-cerebellar system in anthropoids. *Ann. N. Y. Acad. Sci.* 1288:59-69.

Smoot, G. and K. Davidson. 1993. *Wrinkles in Time*. New York: William Morrow & Co.

Socolow, R.H. 1999. Nitrogen management and the future of food: lessons from the management of energy and carbon. *Proc. Natl. Acad. Sci. USA* 96 (11):6001-8.

Srinivasan, M. 2010. Fifteen observations on the structure of energy-minimizing gaits in many simple biped models. *J. Roy. Soc. Interface*: rsif20090544.

Srivatsan, S.G. 2004. Modeling prebiotic catalysis with nucleic acid-like polymers and its implications for the proposed RNA world. *Pure Applied Chem.* 76 (12):2085-2099.

St Clair, J.J.H, and C. Rutz. 2013. New Caledonian crows attend to multiple functional properties of complex tools. *Philos. Trans. R. Soc. B Biol. Sci.* 368 (1630): 20120415.

Stamati, K., V. Mudera, U. Cheema. 2011. Evolution of oxygen utilization in multicellular organisms and implications for cell signaling in tissue engineering. *J. Tissue Engineering* 2:12041731411432365.

Stanley, J., D.R. Loy, and G. Dorje. 2009. *Buddhist Response to Climate Change*. Somerville, MA: Wisdom Publications.

Stearns, S.C. 1977. Evolution of life history traits — critique of theory and review of data. Ann. Rev. Ecol. *Systematics* 8:145-171.

Stiner, M.C., N.D. Munro, T.A. Surovell, E. Tchernov, and O. Bar-Yosef. 1999. Paleolithic population growth pulses evidenced by small animal exploitation. *Science* 283:190-4.

Storz, J. F. 2010. Genes for high altitudes. *Science* 329:40-41.

Strassmann, J. E., and D.C. Queller. 2011. Evolution of cooperation and control of cheating in a social microbe. *Proc. Natl. Acad. Sci. USA* 108 Suppl 2:10855-10862.

Striedter, G.F. 2005. *Principles of Brain Evolution*. 1st ed. Sunderland, MA: Sinauer Associates.

Tangney, J.P., R.F. Baumeister, A.L. Boone. 2004. High self-control predicts good adjustment, less pathology, better grades, and interpersonal success. *J. Personality* 72:271-324.

Tattersal, I. 2014. If I had a hammer. *Sci. Am.* 311 (3):55-59.

Taylor, D.M. and F.M. Moghaddam. 1994. *Theories of intergroup relations: International Social Psychological Perspectives*, 2nd ed. New York: Praeger.

Taylor, E.F. 2003. A call to action. *Am. J. Physics* 71 (5):423-5.

Taylor, S.E. and A.L. Stanton. 2007. Coping resources, coping processes, and mental health. *Ann. Rev Clin. Psychol.* 3:377-401.

Tedeschi, R.G. and L.G. Calhoun. 2004. Posttraumatic growth: Conceptual foundations and empirical evidence. *Psychol. Inquiry* 15:1-18.

Tellis, G.J., E. Yin, S. Bell. 2009. Global consumer innovativeness: cross-country differences and demographic commonalities. *J. Intl. Marketing* 17:1-22.

Tessera, M. and G.A. Hoelzer. 2013. On the thermodynamics of multilevel evolution. *Biosystems* 113 (3):140-3.

Testart, A. 1982. The significance of food storage among hunter-gatherers: residence patterns, population densities, and social inequalities. *Curr. Anthropol.* 23:523-537.

Tetlock, P.E. 1983. Accountability and complexity of thought. *J. Personal. Soc. Psychol.* 45:74-8.

Thalmann, O., B. Shapiro, P. Cui, V.J. Schuenemann, S.K. Sawyer, D.L. Greenfield, M.B. Germonpre et al., 2013. Complete mitochondrial genomes of ancient canids suggest a European origin of domestic dogs. *Science* 342 (6160):871-4.

Tillery, B.W. 2002. *Physical Science*, 5th ed. New York: McGraw-Hill.

Tomasello, M. 2008. *Origins of Human Communication*. Cambridge, MA: MIT Press.

Tomasello, M., A.P. Melis, C. Tennie, E. Wyman, E. Herrmann. 2012 Two key steps in the evolution of human cooperation. *Curr. Anthropol.* 53:673-692.

Tomasello, M. and A. Vaish. 2013. Origins of human cooperation and morality. *Ann. Rev. Psychol.* 64:231-255.

Trefil, J.S., H.J. Morowitz, E. Smith. 2009. The Origin of Life. *Amer. Sci.* 97:206-213.

Trefil, J.S, and R.M. Hazen. 2001. *The Sciences: An Integrated Approach*, 3rd ed. New York: John Wiley & Sons.

Trewavas, A. 2009. What is plant behaviour? *Plant Cell Environ.* 32 (6):606-16.

Trifonova, E.N. and I.N. Berezovsk,Y. 2002. Molecular evolution from abiotic scratch. *Fed. Eur. Biochem. Soc. Lett.* 527:1-4.

Turian, G. 1999. Origin of life. II. From prebiotic replicators to protocells. *Arch. Sci. Compte Rendu Seances Soc.* 52 (2):101-9.

Unwin, G. 2014. History of publishing. *Encyclopedia Britannica Online*, http://www.britannica.com/EBchecked/topic/482597/history-of-publishing.

Uroz, S., P. Oger, C. Lepleux, C. Collignon, P. Frey-Klett, M. P. Turpault. 2011. Bacterial weathering and its contribution to nutrient cycling in temperate forest ecosystems. *Res. Microbiol.* 162 (9):820-831.

Vaccaro, J.A. and S.M Barnett. 2011. Information erasure without an energy cost. *Proc. Royal Soc. A: Math Phys. Eng. Sci.* 467:1770-8.

Valentine, J.W. 1994. Late Precambrian bilaterians: grades and clades. *Proc. Natl. Acad. Sci. USA* 91:6751-7.

Van Boven, L. and L. Ashworth. 2007. Looking forward, looking back: Anticipation is more evocative than retrospection. *J. Exp. Psychol.* 136:289-300.

Vermeij, G. J. 2004. *Nature: An Economic History*. Princeton, NJ: Princeton University Press.

Victor, P.A. 2008. *Managing Without Growth: Slower by Design, Not Disaster*. Cheltenham, U.K.: Edward Elgar Publishing.

Wächtershäuser, G. 1988. Before enzymes and templates: a theory of surface metabolism. *Microb. Rev.* 52:452-484.

Wang, Z., B. Y. Liao, J. Zhang. 2010. Genomic patterns of pleiotropy and the evolution of complexity. *Proc. Natl. Acad. Sci. USA* 107:18034-9.

Ward, C.V., M.W. Tocheri, J.M. Plavcan, F.H. Brown, F.K. Manthi. 2014. Early Pleistocene third metacarpal from Kenya and the evolution of modern human-like hand morphology. *Proc. Natl. Acad. Sci. USA* 111:121-4.

Ward, P. 2009. *The Medea Hypothesis: Is Life on Earth Ultimately Self-Destructive?* Princeton, NJ: Princeton University Press.

Watkins, E.R. 2008. Constructive and unconstructive repetitive thought. *Psychol. Bull.* 134 (2):163-206.

Watkins, E.R. and M. Mounds. 2004. Comparisons between rumination and worry in a non-clinical population. *Behav. Res. Therapy* 43:1577-1585.

Westoff, C.F., and D.A. Koffman. 2011. The association of television and radio with reproductive behavior. *Popul. Devlp. Rev.* 37:749-759.

Whishaw, I.Q., G.A. Metz, B. Kolb, S.M. Pellis. 2001. Accelerated nervous system development contributes to behavioral efficiency in the laboratory mouse: a behavioral review and theoretical proposal. *Dev. Psychobiol.* 39 (3):151-170.

Whitehouse, H. 2004. *Modes of Religiosity: A Cognitive Theory of Religious Transmission.* Lanham, MD: AltaMira Press.

Wicken, J.S. 1979. The generation of complexity in evolution: a thermodynamic and information-theoretical discussion. *J. Theor. Biol.* 77 (3):349-365.

Wicken, J.S. 1985. Thermodynamics and the conceptual structure of evolutionary theory. *J. Theor. Biol.* 117 (3):363-383.

Wilkinson, M. 2010. *Changing Minds in Therapy: Emotion, Attachment, Trauma and Neurobiology.* New York: W.W. Norton.

Wilson, E.O. and B. Holldobler. 2005. Eusociality: origin and consequences. *Proc. Natl. Acad. Sci. USA* 102:13367-13371.

Wilson, E.O. 1999. *Consilience: The Unity of Knowledge.* New York: Vintage Books.

Wilson, M.L. and R.W. Wrangham. 2003. Intergroup relations in chimpanzees. *Ann. Rev. Anthropol.* 32:363-392.

Wisner, B. 2010. Untapped potential of the world's religious communities for disaster reduction in an age of accelerated climate change: An epilogue & prologue. *Religion* 40:128-131.

Wobber, V., B. Hare, R. Wrangham. 2008. Great apes prefer cooked food. *J. Human Evol.* 55:340-8.

Wolszczan, A. 1994. Confirmation of Earth-Mass planets orbiting the millisecond pulsar Psr B1257+12. *Science* 264:538-542.

Wood, B. 1996. Human evolution. *Bioessays* 18 (12):945-954.

Wrangham, R., D. Cheney, R. Seyfarth, E. Sarmiento. 2009. Shallow-water habitats as sources of fallback foods for hominins. *Am. J. Phys. Anthropol.* 140 (4):630-642.

Wrangham, R. and N. Conklin-Brittain. 2003. Cooking as a biological trait. *Comp. Biochem. Physiol. A Mol. Integr. Physiol.* 136:35-46.

Wright, R. 2000. *Nonzero Sum: The Logic of Human Destiny.* New York: Pantheon.

Wright, S. 1932. The roles of mutation, inbreeding, crossbreeding, and selection in evolution. Paper read at Proceedings of the Sixth International Congress of Genetics.

Wurtz, P. and A. Annila. 2010. Ecological succession as an energy dispersal process. *Biosystems* 100 (1):70-8.

Yonge, C.D. 1854. *The Works of Philo Judea: The Contemporary of Josephus, Translated from the Greek.* London: H.G. Bohn.

Zachos, J., M. Pagani, L. Sloan, E. Thomas, K. Billups. 2001. Trends, rhythms, and aberrations in global climate 65 Ma to present. *Science* 292:686-693.

Zeder, M.A., D. Bradley, E. Emshwiller, B.D. Smith, eds. 2010. *Documenting Domestication: New Genetic and Archeological Paradigms.* Oakland CA: University of California Press.

Zipf, G.K. 1949. *Human Behavior and the Principle of Least Effort.* Oxford, UK: Addison-Wesley.

Index

role of writing in 89
Hume, David 99
hydrogen 12, 20, 22, 31, 35, 40, 42-3, 70, 122
hypothalamus 96

I

illness 111-12
industrialization 13, 38, 81, 87-8, 95-6, 126, 140
information
    behavioral uses of 72
    on cosmic scale 71
    and energy 15-16, 30, 36, 50, 69, 81
    social uses of 73
information processing
    about environment 56-7, 71
    to dissipate energy 71
    as measure of complexity 69
    neural
        bioelectrical 57
        origins of 57
        role of neurons in 57
    transition from energy processing 69
    as valued goal 10
insects 78

J

Jurassic Period 47-8

K

K-T boundary 47-8

L

labor
    contract 126
    division of 82, 87, 95, 100
    manual 88, 121
language
    animal 81
    human 81, 83, 86, 89, 100, 136
    origin of 81
    programming 92
learning
    as complement to evolution 72, 106
    conditioning
        classical 72
        instrumental 106, 137, 139, 147, 150
        operant 72
    definition of 72
    habituation 72, 75, 148
    by humans 73, 107-8, 132
    role of 72, 75, 82, 84
    sensitization 72, 75
life
    definition of 33
    evolution of 9, 31, 37, 39, 41, 46, 51, 71, 77, 112, 115, 117, 150
        major transitions in 40
        by natural selection 9, 39-40, 47, 49, 52, 64-5, 82, 96, 106, 109, 116
    history of 10, 40-1, 43-4, 47, 49, 51-2, 70, 77, 100, 109, 138
    meaning of 112
    origin of 34, 141, 144, 147, 151
    on other worlds 114-15
lithium 20, 22, 31, 70

M

mammals 106, 143, 145
mammoths 84-5, 150
Margulis, Lynn 44, 54, 117, 145, 147
Mars 22-3, 26-8, 30, 53
McShea, Daniel 47, 54, 109, 117-18, 146
Medea 107, 116, 152
medulla 63
membrane 41, 43-4, 54, 57, 66
Mercury 22
metabolism
    oxidative 35, 37, 42-6
    rate of 137
    reductive 35, 37, 42
metacognition 72-3, 75, 144, 146
meteorites 53
meteors 24, 31
methane 22, 35, 40, 95
midbrain 61-2
Milky Way galaxy 21-2, 114, 118, 144
mitochondria 44, 108
moons 31
morality 98-9, 113, 131, 142, 147, 149
mutualism 74, 76, 137

N

NADH 40, 53
NADPH 42, 53
natural selection 5, 8-9, 38-40, 43-5, 47, 49-53, 55-6, 64-6, 71-2, 96-7, 99, 105-6, 109-10, 116, 141-4
neolithic 74-5, 79, 118, 142-3, 148
nervous systems
    annelid 58
    arthropod 58
    cephalopod 59
    cnidarian 58
    evolution